The American Commonwealth

The American Commonwealth

–1976–

EDITED BY

Nathan Glazer and Irving Kristol

Basic Books, Inc., Publishers

NEW YORK

CONTENTS

THE AUTHORS

DANIEL PATRICK MOYNIHAN, recent United States Ambassador to the United Nations, has returned to Harvard, where he is a professor in the Department of Government.

SAMUEL P. HUNTINGTON is Frank C. Thomson Professor of Government at Harvard and coeditor of *Foreign Policy*.

MARTIN DIAMOND is Professor of Political Science at Northern Illinois University. His essay was written while he was on a fellowship at the Woodrow Wilson International Center for Scholars in Washington, D.C.

AARON WILDAVSKY is Dean of the Graduate School of Public Policy at Berkeley.

JAMES Q. WILSON is Henry Lee Shattuck Professor of Government at Harvard.

NATHAN GLAZER is Professor of Education and Social Structure at Harvard, and author of *Affirmative Discrimination* (Basic Books, 1975).

IRVING KRISTOL is Henry R. Luce Professor of Urban Values at New York University.

SEYMOUR MARTIN LIPSET is Professor of Government and Sociology at Harvard.

ROBERT NISBET is Albert Schweitzer Professor in the Humanities at Columbia.

DANIEL BELL is Professor of Sociology at Harvard, and author of *The Cultural Contradictions of Capitalism* (Basic Books, 1976).

The American Commonwealth

The American Experiment

DANIEL P. MOYNIHAN

What have we learned? It is two centuries now since the American people commenced what they very well understood to be an experiment in liberty. In his essay in this volume, Martin Diamond, following the lead of Leo Strauss, observes that the men of the Declaration of Independence and of the Constitution were pursuing, were implementing what Hamilton called the "new science of politics." There had been a crucial turn in political thought away from the earlier Greek assumption that the virtue of its citizens was the foundation of the state, and that society accordingly ought to focus on the inculcation of such virtue, and, of course, its further elucidation. Civic philosophy had been the basic political science of such a society, with its teaching a kind of applied science. Such was the view that persisted thereafter for two thousand years, with generally indifferent results. Then came Locke and Montesquieu and others with a quite different view. They saw the object of society as the attainment of liberty for the individual, and judged that society accordingly ought to focus on the practicable arrangements that would establish and preserve that liberty. Diamond describes the Declaration as expressing the "political science of liberty" of that age, a science subsequently, and for the first time, fully elaborated in the Constitution.

Very well, then, what have we learned? Not long ago a friend of many years came through Cambridge, where he presented a paper on Aeschylus to the classics department at Harvard. By the standards of most disciplines, he had chosen late to become a classicist. Something wholly personal had driven him back to the Greeks, and he had stayed there many years and was only just reappearing, in a sense, among his friends. I sought him out and asked the necessary question: What had *he* learned? He thought a moment and replied: "I have learned that if you do not know your future, you will never understand your past."

Which may be so: It is a Greek idea, surely. As it was also the Greeks who first developed the idea of liberty, and the notion of democracy as the form of government in which liberty might reasonably prosper. (Or more accurately, as I learn, "polity"—democracy technically being a corruption of government by popular will, just as tyranny was seen as a corruption of monarchy, and oligarchy a corruption of aristocracy.) Is it then also the case that there is a fated outcome, a destiny—residing in our stars or in ourselves—which, hidden to us now, also conceals the meaning of the events which have been leading to it? It is a possibility not to be dismissed. Indeed, how would we understand 18th-century New England, were it not for the hypothesis that those various worthies were compulsively at work trying to rig the auguries to persuade them of a heavenly destiny of which they could never be confident? But such an understanding goes against the prevailing temper. We assume that causality runs the other way, which is to say forwards not backwards. The setting in which man lives has been transformed by scientists operating in an experimental mode which tries first this, then that, to see which ways successive combinations of choices lead. What then have we learned from the experimental mode applied to the polity?

To read these essays without benefit of a cautionary Introduction (and this alone justifies the reader's brief diversion) would be to risk gaining the impression that we have learned that the American experiment is not going very well. Neither liberty nor democracy would seem to be prospering—or, in any event, neither would seem to have a future nearly as auspicious as their past. The term "auspicious" is intended both as an echo of my friend's Delphic pessimism, and to suggest a variant on the scientists' discovery that observed events are affected by the act of observation. Seemingly nothing at present brings forth more gloom than the contemplation of the future, of which an epiphenomenon is a far too cheerful view of the former times. The present volume may have been noticeably inclined in this direction by the request made of each contributor that he consider his chosen topic in the light of what may be found on the subject in James Bryce's *The American Commonwealth*.

Now, several things may be said of Lord Bryce's unquestionably great work. The first is that it appeared *not quite* a century ago. In 1888, which is the date of the first Preface, things looked considerably more cheerful than they had a mere dozen years earlier, when the United States Centennial Exposition opened in Philadelphia. The experiment, as it turned its first century, didn't look that promising at all. A dozen years later it looked better. A second point is simply that Bryce was an Englishman, and in this case one disposed to be rather more friendly than we often are to ourselves, and in any event concerned with long-term, as against short-term, issues and trends. The heroic opening paragraph of Chapter I takes us in one kaleidoscopic sweep from the fretful concern of the average American regarding the everyday performance of the polity to the grand perspective of the European observer of the manifest and magnificent American destiny:

> "What do you think of our institutions?" is the question addressed to the European traveller in the United States by every chance acquaintance. The traveller finds the question natural, for if he be an observant man his own mind is full of these institutions. But he asks himself why it should be in America only that he is so interrogated. In England one does not inquire from foreigners, nor even from Americans, their views on the English laws and government; nor does the Englishman on the Continent find Frenchmen or Germans or Italians anxious to have his judgment on their politics. Presently the reason of the difference appears. The institutions of the United States are deemed by inhabitants and admitted by strangers to be a matter of more general interest than those of the not less famous nations of the Old World. They are, or are supposed to be, institutions of a new type. They form, or are supposed to form, a symmetrical whole, capable of being studied and judged all together more profitably than the less perfectly harmonized institutions of older countries. They represent an experiment in the rule of the multitude, tried on a scale unprecedentedly vast, and the results of which every one is concerned to watch. And yet they are something more than an experiment, for they are believed to disclose and display the type of institutions towards which, as by a law of fate, the rest of civilized mankind are forced to move, some with swifter, others with slower, but all with unresting feet.

Now obviously the most important fact about the American experiment almost a century later is that ours evidently are *not* the institutions "towards which, as by a law of fate, the rest of civilized mankind are forced to move. . . ." To the contrary, liberal democracy on the American model increasingly tends to the condition of monarchy in the 19th century: a holdover form of government, one which persists in isolated or peculiar places here and there, and may even serve well enough for special circumstances, but which has simply no relevance to the future. It is where the world was, not where it is going. In this respect American institutions reached their apogee in 1919, following World War I, with the extraordinary international position of Woodrow Wilson, who had an eminence no American leader had achieved before, and none since, an eminence it is difficult to imagine any American will ever achieve again under our present arrangements. Through the long unthreatening 19th century—Bryce had said that we sailed a "summer sea"—the expectation that American institutions would inherit the future had no real challengers, or rather, no self-evident ones. Socialist doctrines were forming in Europe, and the equalitarian challenge to liberalism was gathering strength and conviction, but these did not manifest themselves as forms of government, as institutional arrangements, until much later—indeed, not significantly until 1917, and the success of the Bolshevik faction in Russia. Thus the very world events—World War I and its aftermath—which had brought the seeming triumph of the American experiment, now brought into being a new form and theory of government, which promptly assumed the role of the challenger to the American model, and all too soon the American experiment commenced to acquire the fateful air of a transitional arrangement. Other forms of socialism rose to challenge the totalitarian variant. In Western Europe, especially, democratic socialism took hold, but again as a challenge to the American model. A brief restoration—I fear it will be seen as such—took place following the Second World War, with the establishment of post-colonial regimes throughout the tropics which were almost always democratic in form. (They were modeled on European democracies, but by then democracy had become a norm for Europeans as well.) But the regimes did not prosper. In one after another the experiment failed. In 1975, to all appearances, it failed in India, incomparably the largest and most important experiment of all. Increasingly, democracy is seen as an arrangement peculiar to a handful of North Atlantic countries, plus a few of their colonies, as the Greeks would have understood that term.

There have been worse fates than to be a member of this honorable band, and worse by far than to be the American member of it. The world of the 5th century, B.C., was scarcely democratic, but its most vital culture was—and so it is again today. That was not a rich world, nor is today's, but now, as then, the democratic peoples are the most productive, and of these today, Americans incomparably so. (A Royal Commission has reported that in 1973 the richest one percent of the British populace earned incomes over $13,700; the same year the *median* American family income was $12,965.)

And yet the comparison will not do, for if the Greeks had the highest culture of their age, they *knew* it. If the democracies today do, or if the American democracy in particular does, we don't think it, or we don't think we ought, or we don't think we will, or something. . . . The center has not held very well.

The year 1876 was not the best of times either. The future probably looked no better than the awful immediate past. And yet the symbols of the republic were intact. Thus the Philadelphia Exposition was awash with scientific and technological wonders, but the sensation of the event was A. M. Willard's vast romantic canvas, "The Spirit of '76." It probably would be absurd to suppose that any such response could be evoked by a patriotic painting today. Today, more likely, it would be discovered early on that the work had been commissioned by a chromo-lithography dealer who saw the commercial possibilities, as was indeed the case with Willard. In no one thing has the American civic culture declined more in recent decades than in the symbols of love of country, and of manly or womanly pride in the nation. The flag remains, but little else which is not battered or banal or both. The very effort to evoke patriotism is increasingly associated with the near pathetic bewilderment of badly educated and dubiously motivated lower-middle class "conservatives" confronted with the manifest intellectual and social superiority of the nonpatriotic. Even our sense of peoplehood grows uncertain, as ethnic assertions take their implacable toll on the civic assumption of unity.

And so what is left if so much is gone? Curiously, one answer hearkens more to the tradition of the Greeks whom Locke and Montesquieu displaced, than to the tradition established in their stead. For when all else is gone, virtue remains. I venture that the reader of this volume will find it very much present in the *Public Interest* writers here presented.

Ours has been a curious fate. Launched in a brief era of good feeling (I can recall suggesting *Consensus* as a possible name for the new journal, that being a favorite term of President Johnson, in whose administration I was then serving.) *The Public Interest* was soon sailing anything *but* a summer sea. Yet if it was a new ship, it did not have a green crew. In announcing the venture the editors verily proclaimed themselves to be of middle years—"seasoned by life but still open to the future"—and the proposition was tested as the storms of the 1960s broke. It may be that we took the foul weather rather too well, as all about us canvas tore and cables parted. When something like calm returned, we found ourselves confronted with some annoyance, and even a certain mistrust.

We even had our political label changed. When we set out, almost all the editors and contributors to *The Public Interest* would have described themselves as "liberals." (One of the editors would have described himself as a socialist, settling for "right wing social democrat" but his colleagues would insist that that equalled "liberal.") But before long we began to find ourselves depicted as "conservatives." In time a new term—"neoconservative"—was invented for us (more specifically, for myself, Nathan Glazer, and Daniel Bell) by a socialist writer. *Time* and *Newsweek* took up the term, and that settled our future. Past as well, for that matter. Our first reaction to all this was rather in the tradition of the Illinoisan who, being rid out of town on a rail, said that if it wasn't for the honor of the thing he'd just as soon walk. For we had tended to be against labels, and indeed got our own by arguing against those of others.

There had been an event. Rather, a whole series of occurances which had had the conciseness of an event. In the middle of the 1960s a number of social science findings dealt an unprecedented blow to the operating assumptions on which liberal social programs were then being devised. A number of things were learned, of which the most important was that most of the things we were doing probably wouldn't work, in the sense of benefit being proportionate to cost. This was depressing news, but not in itself alarming. The alarming news, or rather the alarming realization, was that so long as this information remained arcane, and the previous assumptions remained generally in place, the outcomes which "we" now increasingly saw to be predictable and explainable would come with surprise and dismay to others, giving rise to all manner of anxiety and deception. And so—in no very organized way, but with a common sense of the situation—we set ourselves the task of breaking the news. It was not welcomed. Why should it have been? We hadn't welcomed it either. The response, as one would expect, was varied. The dimmer sort of folk felt it was sufficient to blame the messenger, and of these the dimmest, or perhaps merely the most obdurate, determined that the thing to do was to do even more of what wouldn't work. At this point an element almost of coercion entered not only the discourse about public policy, but increasingly public policy itself. And at this point the liberals of *The Public Interest* discovered that they were libertarians also, and, indeed, libertarians first.

And so we remain cheerful about all this for one simple reason —the guiding principle which illuminates the essays in this volume. It is, as Martin Diamond shows in his revelatory essay, that liberty was the first principle of this Republic, that it is the animating principle of the Constitution no less than of the Declaration, that it is this which makes us what we are, be we conservative or liberal or, for that matter, socialist—socialists, that is, of the synod with which we commerce. Plainly, any who espouse this principle too assertively may expect in time to have acquired a conservative air. This may be no more than a sign of the Bicentennial times. But this makes it no less the grandest and most glorious idea man has ever had: To espouse it is virtue itself; to do so with a decent competence is all we have ever aspired to. Those who find these essays depressing in their alarm for liberty may take heart in the fact that it is done with such spirit, and so well. Where else?

1

The
democratic
distemper

SAMUEL P. HUNTINGTON

T HE 1960's witnessed a dramatic
upsurge of democratic fervor in America. The predominant trends
of that decade involved challenges to the authority of established
political, social, and economic institutions, increased popular participation in and control over those institutions, a reaction against
the concentration of power in the executive branch of the federal
government and in favor of the reassertion of the power of Congress and of state and local government, renewed commitment to
the idea of equality on the part of intellectuals and other elites, the
emergence of "public interest" lobbying groups, increased concern
for the rights of (and provision of opportunities for) minorities and
women, and a pervasive criticism of those who possessed or were
even thought to possess excessive power or wealth. The spirit of
protest, the spirit of equality, and the impulse to expose and correct inequities were abroad in the land. The themes of the 1960's
were those of Jacksonian Democracy and the muckraking Progressives; they embodied ideas and beliefs which were deep in the

*This article is an abbreviated version of the author's chapter on the
United States in the report to the Trilateral Commission prepared by
Michel Crozier, Samuel P. Huntington, and Joji Watanuki, entitled* The
Governability of Democracies.

American tradition but which did not usually command the passionate intensity of commitment that they did in the 1960's.

This democratic surge manifested itself in an almost endless variety of ways. Consider, for instance, simply a few examples in terms of the two democratic norms of *participation* and *equality*. Though voting participation actually declined, almost all other forms of political participation significantly increased during the 1950's and 1960's. The Goldwater, McCarthy, Wallace, and McGovern candidacies mobilized unprecedented numbers of volunteer campaign workers. In addition, the Republicans in 1962 and the Democrats subsequently launched a series of major efforts to raise a substantial portion of their campaign funds from large numbers of small contributors.

The 1960's also saw, of course, a marked upswing in other forms of citizen participation, in the form of marches, demonstrations, protest movements, and "cause" organizations (such as Common Cause, Nader groups, environmental groups). The expansion of participation thoughout society was reflected in the markedly higher levels of self-consciousness on the part of blacks, Indians, Chicanos, white ethnic groups, students, and women, all of whom became mobilized and organized in new ways to achieve what they considered to be their appropriate share of "the action" and its rewards. In a similar vein, there was a marked expansion of white-collar unionism and of the readiness and willingness of clerical, technical, and professional employees in public and private bureaucracies to assert themselves and to secure protection for their rights and privileges.

In related fashion, the 1960's also saw a reassertion of equality as a goal in social, economic, and political life. The meaning of equality and the means of achieving it became central subjects of debate in intellectual and policy-oriented circles. What was widely hailed as the major philosophical treatise of the decade—*A Theory of Justice,* by John Rawls—defined justice largely in terms of equality. Differences in wealth and power were viewed with increasing skepticism. The classic issue of equality of opportunity versus equality of results was reopened for debate. This intellectual concern over equality did not, of course, easily transmit itself into widespread reduction of inequality in society. But the dominant thrust in political and social action was clearly in that direction.

Whatever its causes—and books are being written about that—the consequences of the democratic surge of the 1960's will be felt for years to come. The analysis here focuses on the immediate—and somewhat contradictory—effects of the democratic surge on government. The basic point is this: *The vitality of democracy in the United States in the 1960's produced a substantial increase in governmental activity and a substantial decrease in governmental authority.* By the early 1970's Americans were progressively demanding and receiving more benefits from their government and yet having less confidence in their government than they had a decade earlier. And, paradoxically, this working out of the democratic impulse was also associated on the one hand with a decline of the more political, interest-aggregating, "input" institutions of government (most notably, political parties and the Presidency), and on the other hand with the growth in the bureaucratic, regulating, "output" institutions of government. The apparent vitality of democracy in the 1960's raises questions about the governability of democracy in the 1970's.

The expansion of governmental activity

The structure of governmental activity in the United States—in terms of both its size and its content—went through two major changes during the quarter-century after World War II.[1] The first change, the "defense shift," was a response to the external Soviet threat of the 1940's; the second change, the "welfare shift," was a response to the internal democratic surge of the 1960's. The former was primarily the product of elite leadership; the latter was primarily the result of popular expectations and group demands.

In 1948, total governmental expenditures (federal, state, and local) amounted to 20 per cent of GNP; national defense expenditures were four per cent of GNP; and governmental purchases of goods and services were 12 per cent of GNP. During the next five years these figures changed drastically. The changes were almost entirely due to the onslaught of the Cold War and the perception, eventually shared by the top executives of the government—Truman, Acheson, Forrestal, Marshall, Harriman, Lovett—that a major effort was required to counter the Soviet threat to the security of the West. The key turning points in the development of that perception included Soviet pressure on Greece and Turkey, the Czech coup, the Berlin blockade, the Communist conquest of China, the Soviet atomic explosion, and eventually the North Korean attack on South Korea. In late 1949, a plan for major rearmament to meet this threat had been drawn up within the executive branch. The top executive leaders, however, felt that neither Congress nor public opinion was ready to accept such a large-scale military build-up.

These political obstacles were removed by the outbreak of the Korean War in June 1950.

The result was a major expansion in U.S. military forces and a drastic reshaping of the structure of governmental expenditures and activity. By 1953, national defense expenditures had gone up from their 1948 level of four per cent of GNP to over 13 per cent of GNP. Non-defense expenditures remained at about 15 per cent of GNP. The governmental share of the output of the American economy increased by about 80 per cent during these five years—virtually all of it in the national defense sector.

With the advent of the Eisenhower Administration and the end of the Korean War, these proportions shifted somewhat and then settled into a relatively fixed pattern which remained markedly stable for over a decade. From 1954 to 1966, governmental expenditures were usually about 27 or 28 per cent of GNP; government purchases of goods and services varied between 19 and 22 per cent; and defense expenditures, with the exception of a brief dip in 1964 and 1965, were almost constant at nine to 10 per cent of GNP.

In the mid-1960's, however, the stability of this pattern was seriously disrupted. The Vietnam War caused a minor disruption, reversing a downward trend in the defense proportion of GNP and temporarily maintaining defense expenditures at nine per cent of GNP. The more significant and lasting change was the tremendous expansion of the non-defense activities of government. Between 1965 and 1974, total governmental expenditures rose from 27 to 33 per cent of GNP; governmental purchases of goods and services, on the other hand, which also had increased simultaneously with total expenditures between 1948 and 1953, changed only modestly from 20 per cent in 1965 to 22 per cent in 1974. This difference meant, of course, that a substantial proportion of the increase in governmental spending was in the form of transfer payments, e.g., welfare and social security benefits, rather than in the form of additional governmental contributions to the GNP. Non-defense expenditures, which had been 20 per cent of GNP in 1965, were 25 per cent of GNP in 1971 and an estimated 27 per cent of GNP in 1974. Defense spending went down to seven per cent of GNP in 1971 and six per cent in 1974.

In fiscal year 1960, total foreign affairs spending accounted for 53.7 per cent of the federal budget, while expenditures for cash income maintenance accounted for 22.3 per cent. In fiscal year 1974, according to Brookings Institution estimates in *Setting National Priorities: The 1974 Budget,* almost equal amounts were spent for both these purposes, with foreign affairs taking 33 per cent and cash income maintenance 31 per cent of federal budget. Across the board, the tendency was for massive increases in governmental expenditures to provide cash and benefits for particular individuals and groups within our society.

The "welfare shift"

The "welfare shift," like the "defense shift" before it, underlined the close connection between the structure of governmental activity and the trend of public opinion. During the 1940's and early 1950's, the American public willingly approved massive programs for defense and international affairs: When queried as to whether the military budget or the size of the armed forces should be increased, decreased, or remain about the same, the largest proportions of the public almost consistently supported a greater military effort. During the middle and later 1950's, after defense spending had in fact expanded greatly, support for still further expansion eased somewhat. But even then, only a small minority of the public supported a decrease, with the largest group approving the existing level of defense effort. Popular support for other government programs, including all domestic programs and foreign aid, almost always was substantially less than support for defense spending.

During the mid-1960's, at the peak of the democratic surge and of the Vietnam war, public opinion on these issues changed drastically. When asked in 1960, for instance, how they felt about current defense spending, 18 per cent of the public said the United States was spending too much on defense, 21 per cent too little, and 45 per cent said the existing level was about right. Nine years later, in July 1969, the proportion of the public saying that too much was being spent on defense had zoomed up from 18 to 52 per cent; the proportion thinking that too little was being spent on defense had dropped from 21 to eight per cent; the proportion approving the current level had declined from 45 to 31 per cent. This new pattern of opinion on defense remained relatively stable during the late 1960's and early 1970's. Simultaneously, public opinion became more favorable to governmental spending for domestic programs.

After the "defense shift," during the 1950's and early 1960's, governmental expenditures normally exceeded governmental revenue, but with one exception (1959, when the deficit was $15 billion), the gap between the two was not large in any single year. In the late 1960's, on the other hand, after the fiscal implications of the "welfare shift" had been felt, the overall governmental deficit took on new proportions. In 1968 it was $17 billion, and in 1971, $27 billion. The cumulative deficit for the five years from 1968 through 1971 was $43 billion.

The excess of expenditures over revenues was obviously one major source of the inflation which plagued the United States in the early 1970's. Inflation was, in effect, one way of paying for the new forms of government activity produced by the "welfare shift." The extent of the fiscal gap, its apparent inevitability and intractableness, and its potentially destabilizing effects were sufficiently ominous to generate a new variety of Marxist analysis of the inevitable collapse of capitalism. "The fiscal crisis of the capitalist state," according to James O'Connor in *The Fiscal Crisis of the State,* "is the inevitable consequence of the structural gap between state expenditures and revenues." As Daniel Bell has suggested in *The Public Interest* (Fall 1974), in effect this argument represents a version of what can only be called neo-neo-Marxism: The original Marxism said the capitalist crisis would result from anarchical competition; neo-Marxism said it would be the result of war and war-expenditures; now, the most recent revision, taking into consideration the "welfare shift," identifies the expansion of social expenditures as the source of the fiscal crisis of capitalism. What the Marxists mistakenly attribute to capitalist economics, however, is in fact a product of democratic politics.

The "defense shift" involved a major expansion of the national effort devoted to military purposes, followed by slight reduction and stabilization of the relation of that activity to the total national product. The "welfare shift" has produced a comparable expansion and redirection of governmental activity. The key question is: To what extent will this expansion be limited in scope and time, as was the defense expansion, or to what extent will it be an open-ended, continuing phenomenon? Has non-defense governmental spending peaked at about 27 per cent of GNP? These are not easy questions to answer.

On the one hand, history suggests that the recipients of subsidies have more specific interests, are more self-conscious and organized, and are better able to secure access to the political decision points than are the more amorphous, less well-organized, and more diffuse taxpaying interests. On the other hand, there is also some evidence that the conditions favorable to large-scale governmental programs which existed in the 1960's may now be changing significantly. Some polls suggest that the public has become more conservative in its attitudes towards government generally and more hostile towards governmental spending in particular. Still, most observers would agree that a decline in government non-defense spending is extremely unlikely. At the very least, the "welfare shift" has effected a permanent change in the relationship between government and society in the United States.

The decline in governmental authority

The essence of the democratic surge of the 1960's was a general challenge to existing systems of authority, public and private. In one form or another, this challenge manifested itself in the family, the university, business, public and private associations, politics, the governmental bureaucracy, and the military services. People no longer felt the same obligation to obey those whom they had previously considered superior to themselves in age, rank, status, expertise, character, or talents. Within most organizations, discipline eased and differences in status became blurred. Each group claimed its right to participate equally—and perhaps more than equally—in the decisions which affected itself. The questioning of authority pervaded society. In politics, it manifested itself in a decline in public confidence and trust in political leaders and institutions, a reduction in the power and effectiveness of political institutions such as the political parties and the Presidency, a new importance for the "adversary" media and "critical" intelligentsia in public affairs, and a weakening of the coherence, purpose, and self-confidence of political leadership.

The decline in confidence in governmental leaders and institutions was related to a somewhat earlier tendency towards ideological and policy polarization which, in turn, had its roots in the expansion of political participation in the late 1950's and early 1960's. The democratic surge involved a more politically active citizenry, which developed increased ideological consistency on public issues, and then lost its confidence in public institutions and leaders when governmental policies failed to correspond to what it desired. The sequence and direction of these shifts in public opinion dramatically illustrates how the upsurge of democracy in the 1960's (as manifested in increased political participation) produced problems for the governability of democracy in the 1970's (as manifested in the decreased public confidence in government).

During the 1960's public opinion on major issues of public policy tended to become more ideologically structured—that is, people tended to hold more consistent liberal or conservative attitudes on public policy issues. Thus, the image of the American voter as independently and pragmatically making up his mind in ad hoc fashion on "the merits" of different issues became rather far removed from actuality.

This pattern of developing polarization and ideological consistency had its roots in two factors. First, those who are more active in politics are also more likely to have consistent and systematic views on policy issues. The increase in political participation in the early 1960's was thus followed by heightened polarization of political opinion in the mid-1960's. The increase in polarization, in turn, often involved higher levels of group consciousness (as among blacks) which then stimulated more political participation (as in the white backlash).

Second, the polarization was clearly related to the nature of the issues which became the central items on the political agenda of the mid-1960's. The three major clusters of issues which then came to the fore were: social issues, such as the use of drugs, civil liberties, and the role of women; racial issues, involving integration, busing, government aid to minority groups, and urban riots; military issues, involving primarily, of course, the war in Vietnam but also the draft, military spending, military aid programs, and the role of the military-industrial complex more generally. All three sets of issues, but particularly the social and racial issues, tended to generate high correlations between the stands which people took on individual issues and their overall political ideology. On more strictly economic issues, on the other hand, ideology was a much less significant factor. Thus, to predict an individual's position on the legalization of marijuana or school integration or the size of the defense budget, one would want to ask him whether he considered himself a liberal, a moderate, or a conservative. To predict his stand on federally financed health insurance, one should ask him whether he was Democrat, Independent, or Republican.[2]

Polarization and confidence

This polarization over issues in the mid-1960's in part, at least, explains the major decline in trust and confidence in government of the later 1960's. Increasingly, substantial portions of the American public took more extreme positions on policy issues; those who took more extreme positions on policy issues, in turn, tended to become more distrustful of government.[3] Political leaders, in effect, alienated more and more people by attempting to please them through the time-honored traditional politics of compromise.

At the end of the 1950's, for instance, about three quarters of the American people surveyed by the University of Michigan Survey Research Center thought that their government was run primarily for the benefit of the people, and only 17 per cent thought that it primarily responded to what "big interests" wanted. These proportions steadily changed during the 1960's, stabilizing at very different levels in the early 1970's. By the latter half of 1972, only 38 per cent of the population thought that government was "run for the benefit of all the people," and a majority of 53 per cent thought that it was "run by a few big interests looking out for themselves." In 1959, when asked what they were most proud of about their country, 85 per cent of Americans (as compared to 46 per cent of Britons, 30 per cent of Mexicans, seven per cent of Germans, and three per cent of Italians in the same comparative survey) mentioned their "political institutions." By 1973, however, 66 per cent of a national sample of Americans said that they were dissatisfied by the way in which their country was governed.[4] In similar fashion, in 1958, 71 per cent of the population felt that they could trust the government in Washington to do what was right "all" or "most" of the time, while only 23 per cent said that they could trust it only "some" or "none" of the time. By late 1972, however, the number who would trust the national government to do what was right all or most of the time had declined to 52 per cent, while the number who thought it would do what was right only some or none of the time had doubled to 45 per cent. Again, the pattern of change shows a high level of confidence in the 1950's, a sharp decline in confidence during the 1960's, and a leveling off at much reduced levels of confidence in the early 1970's.

The precipitous decline in public confidence in their leaders in the latter part of the 1960's and the leveling off or partial restoration of confidence in the early 1970's can also be seen in other data which permit some comparison between attitudes towards government and towards other major institutions in society.[5] Between 1966 and 1971 the proportion of the population having a "great deal of confidence" in the leaders of each of the major governmental institutions was cut in half. By 1973, however, public confidence in the leadership of the Congress, the Supreme Court, and the military had begun to be renewed from the lows of two years earlier. Confidence in the leadership of the executive branch, on the other hand—not surprisingly—was at its lowest point. These changes of attitudes toward governmental leadership did not occur in a vacuum, but were part of a general weakening of confidence in institutional leadership. The leadership of the major non-governmental institutions in society which had enjoyed high levels of public confidence in the mid-1960's—large corporations, higher educational institutions, and medicine—also suffered a somewhat similar pattern of substantial decline and partial recovery. Significantly, only the leadership of the press and television news enjoyed more confidence in 1973 than it had in 1966, and only in the case of television was the increase a substantial and dramatic one. In 1973 the institutional leaders in which the public had the greatest degree of confidence were, in declining order of confidence: medicine, higher education, television news, and the military.

The late 1960's and early 1970's also saw a significant decline from the levels of the mid-1960's in the sense of political efficacy on the part of large numbers of people. In 1966, for instance, 37 per cent of the people believed that what they thought "doesn't count much anymore"; in 1973, a substantial majority of 61 per cent of the people believed this. Similarly, in 1960, 42 per cent of the American public scored "high" on a "political efficacy index" developed by the Michigan Survey Research Center and only 28 per cent of the population scored "low." By 1968, however, this distribution had changed dramatically, with 38 per cent of the people scoring "high" and 44 per cent of the population scoring "low."

The logic of political participation

In terms of traditional theory about the requisites for a viable democratic policy, these trends of the 1960's thus end up as a predominantly negative, but still mixed, report. On the one side, there are the increasing distrust and loss of confidence in government, the tendencies towards the polarization of opinion, and the declining sense of political efficacy. On the other side, there is the early rise in political participation over previous levels. As we have suggested, these various trends all may be interrelated. The increases in participation first occurred in the 1950's; these were followed by the polarization over racial, social, and military issues in the mid-1960's; this, in turn, was followed by the decrease in confidence and trust in government and in one's own political efficacy in the late 1960's. There is reason to believe, as Sidney Verba has suggested, that this sequence was not entirely accidental. Those who are active in politics are likely to have more systematic and consistent views on political issues; and those who have such views are, as we have seen, likely to become alienated if government action does not reflect their views. This logic also would suggest that those who are most active politically should be most dissatisfied with the political system. In the past, exactly the reverse has been the case: The active political participants have highly positive attitudes towards government and politics. Now, however, this relationship seems to be weakening.

The decline in the average citizen's sense of political efficacy could also produce a decline in levels of political participation. If this should be the case, one might well think of a cyclical process of interaction in which: 1) political participation leads to increased policy polarization within society; 2) increased policy polarization leads to increasing distrust and a decreasing sense of political efficacy among individuals; 3) a decreasing sense of political efficacy leads to decreased political participation.

In addition, change in the principal issues on the political agenda could lead to decreasing ideological polarization. The concern with many of the heated issues of the 1960's has been displaced on the public agenda by an overwhelming preoccupation with economic issues, first inflation and then recession and unemployment. The positions of people on economic issues, however, are not as directly related to their basic ideological inclinations as their positions on other issues. In addition, inflation and unemployment are like crime; no one is in favor of them, and significant differences can only appear if there are significantly different alternative programs for dealing with the problem. Such programs, however, have been slow in materializing, and hence the salience of economic issues may give rise to generalized feelings of lack of confidence in the political system—but not to dissatisfaction rooted in the failure of government to follow a particular set of policies. Such generalized alienation could, in turn, reinforce tendencies towards political passivity engendered by the already observable decline in the sense of one's own political efficacy. This suggests that the democratic surge of the 1960's could well be generating its own countervailing force.

The decay of the party system

The decline in the role of political parties in the United States in the 1960's can be seen in a variety of ways.

1. Party identification has dropped sharply, and the proportion

of the public which considers itself "Independent" in politics has correspondingly increased. In 1972 more people identified themselves as Independent than identified themselves as Republican, and among those under age 30 there were more Independents than Republicans *and* Democrats combined. Younger voters always tend to be less partisan than older voters. But the proportion of Independents among this age group has gone up sharply. In 1950, for instance, 28 per cent of the 21-to-29-year-old group identified themselves as Independent; in 1971, 43 per cent of this age group did. Thus, unless there is a reversal of this trend, substantially lower levels of party identification among the American electorate are bound to persist for at least another generation.

2. Party voting has declined, and ticket-splitting has become a major phenomenon. In 1950 about 80 per cent of the voters cast straight party ballots; in 1970 only 50 per cent did. Voters are thus more inclined to vote the man than to vote the party; and this, in turn, means that each candidate has to campaign primarily as an individual and sell himself to the voters in terms of his own personality and talents, rather than joining with the other candidates of his party in a collaborative partisan effort. Hence he must also raise his own money and create his own organization. The phenomenon, represented at the extreme by CREEP and its isolation from the Republican National Committee in the 1972 election, is being duplicated in greater or lesser measure in most other electoral contests.

3. Partisan consistency in voting is also decreasing—that is, voters are more likely to vote Republican in one election and Democratic in the next. At the national level, there is a growing tendency for public opinion to swing generally back and forth across the board, with relatively little regard to the usual differences among categorical voting groups. Four out of the six Presidential elections since 1952 have been landslides. This phenomenon is a product of the weakening of party ties and the decline of regionalism in politics. In 1920, Harding received about the same percentage of the popular vote as Nixon did in 1972, but Harding lost 11 states while Nixon lost only Massachusetts and the District of Columbia.[6] In a similar vein, the fact that the voters cast 60 per cent of their votes for Nixon in 1972 did not prevent them from reelecting a Democratic Congress that year and then giving the Democrats an even more overwhelming majority in Congress two years later.

As the above figures suggest, the significance of party as a guide to electoral behavior has declined substantially. In part, but only in part, candidate appeal took its place. Even more important was the rise of issues as a significant factor affecting voting behavior. Previously, if one wanted to predict how an individual would vote in a Congressional or Presidential election, the most important fact to know about him was his party identification. This is no longer the case. In 1956 and 1960, party identification was three or four times as important as the views of the voter on issues in predicting how he would vote; but in 1964 and in subsequent elections, this relationship reversed itself. *Issue politics has replaced party politics as the primary influence on mass political behavior.*[7] In addition, party identification is no longer as significant a guide as it once was to how Congressmen will vote. In the House of Representatives, for instance, during Truman's second term, 54.5 per cent of the roll call votes were party unity votes, in which a majority of one party opposed a majority of the other party. This proportion has declined steadily, to the point where in Nixon's first term only 31 per cent of the roll call votes were party unity votes.[8]

Party decomposition

Not only has the mass base of the parties declined but so also has the coherence and strength of party organization. The political party has, indeed, become less of an organization, with a life and interest of its own, and more of an arena in which other actors pursue their interests. In some respects, of course, the decline of party organization is an old and familiar phenomenon. The expansion of government welfare functions beginning with the New Deal, the increased pervasiveness of the mass media, particularly television, and the higher levels of affluence and education among the public all have tended over the years to weaken the traditional bases of party organization. During the 1960's, however, this trend seemed to accelerate. In both major parties, party leaders found it difficult to maintain control of the most central and important function of the party: the selection of candidates for public office. In part, the encroachment on party organization was the result of the mobilization of "issue-constituencies" by issue-oriented candidates, of whom Goldwater, McCarthy, Wallace, and McGovern were the principal examples on the national level. In part, however, it was simply a reaction against party politics and party leaders. Endorsements by party leaders or by party conventions carried little positive weight and often were a liability. The "outsider" in politics, or the candidate who could make himself appear to be an outsider, now had the inside road to political office. In New York in 1974, for instance,

four of five candidates for state-wide office endorsed by the state
Democratic convention were defeated by the voters in the Democratic primary; the party leaders, it has been aptly said, did not
endorse Hugh Carey for governor because he could not win, and
he won because they did not endorse him.

The trends in party reform and organization in the 1960's were
all designed to "open the parties" and to encourage fuller participation in party affairs. In some measure, these reforms conceivably
could mitigate a peculiar paradox in which popular participation in
politics was going up, but the premier organization designed to
structure and organize that participation—the political party—was
declining. At the same time, the longer-term effect of the reforms
could be very different from that which was intended. In the democratic surge during the Progressive Era at the turn of the century,
the direct primary was widely adopted as a means of insuring popular control of the party organization. In fact, however, the primary
reinforced the power of the political bosses whose followers in the
party machine always voted in the primaries. In similar fashion,
the reforms within the Democratic Party to insure the representation of all significant groups and viewpoints in party conventions
appear likely to give the party leaders—skilled at compromise and
political brokerage—new influence over the choice of the Presidential nominee at the next national convention.

It is true that the signs of decay in the American party system
have their parallels in the party systems of other industrialized democratic countries. In addition, however, the developments of the
1960's in the American party system can also be viewed in terms
of the historical dynamics of American politics. According to the
standard theory of American politics, a major party realignment occurs, usually in conjunction with a "critical election," approximately
every 28 to 36 years: 1800, 1828, 1860, 1896, 1932. In terms of this
theory, such a realignment was obviously due about 1968. In fact,
many of the signs of party decay which were present in the 1960's
historically also have accompanied major party realignments: a decline in party identification, increased electoral volatility, third-party
movements, the loosening of the bonds between social groups and
political parties, and the rise of new policy issues which cut across
the older cleavages. The decay of the old New Deal party system
was clearly visible, and the emergence of a new party system was
eagerly awaited, at least by politicians and political analysts. Yet
neither in 1968 nor in 1972 did a new coalition of groups emerge
to constitute a new partisan majority and give birth to a new party

alignment. Nor did there seem to be any significant evidence that such a realignment was likely in 1976—by which time it would be eight to 16 years "overdue," according to the "normal" pattern of party system evolution.

Alternatively, the signs of party decomposition could be interpreted as presaging not simply a realignment of parties within an ongoing system but rather a more fundamental decay and potential dissolution of the party system. In this respect, it could be argued that the American party system emerged during the Jacksonian years of the mid-19th century, that it went through realignments in the 1850's, 1890's, and 1930's, but that it had reached its peak (in terms of popular commitment and organizational strength) in the last decades of the 19th century, and that since then it has been going through a slow, but now accelerating, process of disintegration. To support this proposition, it could be argued that political parties are a political form peculiarly suited to the needs of industrial society and that the movement of the United States into a "post-industrial" phase means the end of the political party system as we have known it. If this be the case, a question will have to be faced: Is democratic government possible without political parties? And if so, how?

Government and opposition: the shifting balance

The governability of a democracy depends upon the relation between the authority of its governing institutions and the power of its opposition institutions. In a parliamentary system, the authority of the cabinet depends upon the balance of power between the governing parties and the opposition parties in the legislature. In the United States, the authority of government depends upon the balance of power between a broad coalition of governing institutions and groups, which includes but transcends the executive and other formal institutions of government, and those institutions and groups which are committed to the opposition. During the 1960's and early 1970's the balance of power between government and opposition shifted significantly. The central governing institution in the political system, the Presidency, declined in power; institutions playing opposition roles in the system, most notably the national media and Congress, significantly increased their power.

"Who governs?" is obviously one of the most important questions to ask concerning any political system. Even more important, however, may be the question, "Does anybody govern?" To the extent that the United States was governed by anyone during the decades after World War II, it was governed by the President acting with the support and cooperation of key individuals and groups in the executive office, the federal bureaucracy, Congress, and the more important businesses, banks, law firms, foundations, and media, which constitute the private sector's "Establishment." In the 20th century, whenever the American political system has moved systematically with respect to public policy, the direction and the initiative have come from the White House. When the President is unable to exercise authority, when he is unable to command the cooperation of key decision-makers elsewhere in society and government, no one else has been able to supply comparable purpose and initiative. To the extent that the United States has been governed on a national basis, it has been governed by the President. During the 1960's and early 1970's, however, the authority of the President declined significantly, and the governing coalition which had, in effect, helped the President to run the country from the early 1940's down to the early 1960's began to disintegrate.

These developments were, in some measure, a result of the extent to which all forms of leadership, but particularly those associated with or tainted by politics, tended to lose legitimacy in the 1960's and 1970's. Not only was there a decline in the confidence of the public in political leaders, but there was also a marked decline in the confidence of political leaders in themselves. In part, this was the result of what were perceived to be significant policy failures: the failure to "win" the war in Indochina, the failure of the Great Society's social programs to achieve their anticipated results, and the intractibility of inflation. These perceived failures induced doubts among political leaders as to the effectiveness of their rule. In addition, and probably more important, political leaders also had doubts about the morality of their rule. They too shared in the democratic, participatory, and egalitarian ethos of the times; and hence they too had questions about the legitimacy of hierarchy, coercion, discipline, secrecy, and deception—all of which are, in some measure, inescapable attributes of the process of government.

The decline of the Presidency

Probably no development of the 1960's and 1970's has greater import for the future of American politics than the decline in the authority, status, influence, and effectiveness of the Presidency. The effects of the weakening of the Presidency will be felt throughout the body politic for years to come. This decline of the Presidency manifests itself in a variety of ways.

No one of the last four Presidents has served a full course of eight years in office. One President has been assassinated, one was forced out of office because of opposition to his policies, and another was forced out because of personal opposition. Short terms in office reduce the effectiveness of the President in dealing with enemies and allies abroad and bureaucrats and Congressmen at home. The greatest weakness in the Presidency in American history was during the period from 1848 to 1860—12 years in which four men occupied the office, and none of them was reelected.

At present, for the first time since the Jacksonian Revolution, the United States has a President and a Vice President who are not the product of a national electoral process. Both the legitimacy and the power of the Presidency are weakened to the extent that the President does not come into office through an involvement in national politics which compels him to mobilize support throughout the country, to negotiate alliances with diverse economic, ethnic, and regional groups, and to defeat his adversaries in intensely fought state and national electoral battles. The current President is a product of Grand Rapids and the House—not of the nation. The United States has almost returned, at least temporarily, to the relations between Congress and President which prevailed during the Congressional caucus period in the second decade of the 19th century.

Since Theodore Roosevelt, at least, the Presidency has been viewed as the most popular branch of government and that which is most likely to provide the leadership for progressive reform. Liberals, progressives, and intellectuals have all seen the Presidency as the key to change in American politics, economics, and society. The great Presidents have been the strong Presidents, who stretched the legal authority and political resources of the office to mobilize support for their policies and to put through their legislative program. In the 1960's, however, the tide of opinion dramatically reserved itself: Those who previously glorified Presidential leadership now warn of the dangers of Presidential power.

While much was made in the press and elsewhere during the 1960's about the dangers of the abuses of Presidential power, this criticism of Presidential power was, in many respects, evidence of the decline of Presidential power. Certainly the image which both Presidents Johnson and Nixon had of their power was far different—and probably more accurate, if only because it was self-

fulfilling—than the images which the critics of the Presidency had
of Presidential power. Both Johnson and Nixon saw themselves
as isolated and beleaguered, surrounded by hostile forces in the
bureaucracy and the Establishment. Under both of them, a feeling of almost political paranoia pervaded the White House: a
sense that the President and his staff were an "island" in a hostile world. Under Nixon, these feelings of suspicion and mistrust led members of the President's staff to engage in reckless, illegal, and self-defeating actions to counter his "enemies"; at the same
time, these feelings also made it more difficult for them to engage
in political compromises and to exercise political leadership, both
of which are necessary to mobilize supporters.

During the late 1960's and early 1970's, Congress and the courts
began to impose a variety of formal restrictions on Presidential
power: the War Powers Act, the budgetary reform act, the limits on Presidential impoundment of funds, and similar measures.
At the same time, and more important, the effectiveness of the
President as the principal leader of the nation also declined as
a result of the diminished effectiveness of leadership at other levels
in society and government. The absence of strong central leadership in Congress (on the Rayburn-Johnson model, for instance)
made it impossible for a President to secure support from Congress
in an economical fashion. The diffusion of authority in Congress
meant a reduction in the authority of the President. There was no
central leadership with whom he could negotiate and come to terms.
The same was true with respect to the Cabinet. The general decline
in the status of Cabinet Secretaries was often cited as evidence of
the growth in the power of the Presidency on the grounds that
the White House staff was assuming powers which previously
rested with the Cabinet. But in fact the decline in the status of
Cabinet Secretaries made it more difficult for the President to command the support and cooperation of the executive bureaucracy;
weak leadership at the departmental level produces weakened leadership at the President level.

Electoral coalitions and governing coalitions

To become President a candidate has to put together an *electoral
coalition* involving a majority of voters appropriately distributed
across the country. He normally does this by 1) developing an identification with certain issues and positions which bring him the support of key categorical groups—economic, regional, ethnic, racial,

and religious; and 2) cultivating the appearance of certain general characteristics—honesty, energy, practicality, decisiveness, sincerity, and experience—which appeal across the board to people in all categorical groups. Before the New Deal, when the needs of the national government in terms of policies, programs, and personnel were relatively small, the President normally relied on the members of his electoral coalition to help him govern the country. Political leaders in Congress, in the state houses, and elsewhere across the country showed up in Washington to man the administration, and the groups which comprised the electoral coalition acted to put through Congress the measures in which they were interested.

Since the 1930's, however, the demands on government have grown tremendously and the problems of constituting a *governing coalition* have multiplied commensurately. Indeed, once he is elected President, the President's electoral coalition has, in a sense, served its purpose. The day after his election the size of his majority is almost—if not entirely—irrelevant to his ability to govern the country. What counts then is his ability to mobilize support from the leaders of the key institutions in society and government. He has to constitute a broad governing coalition of strategically located supporters who can furnish him with the information, talent, expertise, manpower, publicity, arguments, and political support which he needs to develop a program, to embody it in legislation, and to see it effectively implemented. This coalition must include key people in Congress, the executive branch, and the private-sector "Establishment." The governing coalition need have little relation to the electoral colation. The fact that the President as a candidate put together a successful electoral coalition does not insure that he will have a viable governing coalition.

For 20 years after World War II, Presidents operated with the cooperation of a series of informal governing coalitions. Truman made a point of bringing a substantial number of non-partisan soldiers, Republican bankers, and Wall Street lawyers into his Administration. He went to the existing sources of power in the country to get the help he needed in ruling the country. Eisenhower in part inherited this coalition and was in part almost its creation. He also mobilized a substantial number of Midwestern businessmen into his Administration and established close and effective working relationships with the Democratic leadership of Congress. During his brief Administration, Kennedy attempted to recreate a somewhat similar structure of alliances. Johnson was acutely aware of the need to maintain effective working relations with the "Eastern Establishment" and other key groups in the private sector, but, in effect, in 1965 and 1966 was successful only with respect to Congress. The informal coalition of individuals and groups which had buttressed the power of the three previous Presidents began to disintegrate.

Both Johnson and his successor were viewed with a certain degree of suspicion by many of the more liberal and intellectual elements which normally might have contributed support to their Administrations. The Vietnam War and, to a lesser degree, racial issues divided elite groups as well as the mass public. In addition, the number and variety of groups whose support might be necessary had increased tremendously by the 1960's. Truman had been able to govern the country with the cooperation of a relatively small number of Wall Street lawyers and bankers. By the mid-1960's, the sources of power in society had diversified tremendously, and this was no longer possible.

The new role of the media

The most notable new source of national power in 1970, as compared to 1950, was the national media—meaning the national television networks, the national news magazines, and the major newspapers with national reach, such as the *Washington Post* and the *New York Times*. It is a long-established and familiar political fact that within a city, and even within a state, the power of the local press serves as a major check on the power of the local government. In the early 20th century, the United States developed an effective national government, making and implementing national policies. Only in recent years, however, has there come into existence a national press with the economic independence and communications reach to play a role with respect to the President that a local newspaper plays with respect to a mayor. This marks the emergence of a very significant check on Presidential power. In the two most dramatic domestic policy conflicts of the Nixon Administration—the Pentagon Papers and Watergate—organs of the national media challenged and defeated the national executive. The press, indeed, played a leading role in bringing about what no other single institution, group, or combination of institutions and groups had done previously in American history: Forcing out of office a President who had been elected less than two years before by an overwhelming popular majority. No future President can or will forget that fact.

The late 1960's and early 1970's also saw a reassertion of the power of Congress. In part, this represented simply the latest phase in the institutionalized constitutional conflict between Congress and President; in part, also, of course, it reflected the fact that after 1968, the Presidency and the Congress were controlled by different parties. In addition, however, these years saw the emergence, first in the Senate and then in the House, of a new generation of Congressional activists willing to challenge established authority in their own chambers as well as in the executive branch.

The increased power of the national opposition, centered in the press and in Congress, undoubtedly is related to and perhaps is a significant cause of the critical attitudes which the public has towards federal, as compared to state and local, government. While data for past periods are not readily available, certainly the impression one gets is that over the years the public often has tended to view state and local government as inefficient, corrupt, inactive, and unresponsive. The federal government, on the other hand, has seemed to command much greater confidence and trust, going all the way from early childhood images of the "goodness" of the President to respect for the Federal Bureau of Investigation, Internal Revenue Service, and other federal agencies having an impact on the population as models of efficiency and integrity. It now would appear that there has been a drastic reversal of this confidence. In 1973, a national sample was asked whether it then had more or less confidence in each of the three levels of government than it had had five years previously. Confidence in all three levels of government declined more than it rose, but the proportion of the public which reported a decline in confidence in the federal government (57 per cent) was far higher than that which reported a decline in confidence in state (26 per cent) or local (30 per cent) government. As one would expect, substantial majorities also went on record in favor of increasing the power of state government (59 per cent) and of local government (61 per cent). But only 32 per cent wanted to increase the power of the federal government, while 42 per cent voted to decrease its power.[9]

The balance between government and opposition depends not only on the relative power of different institutions, but also on their roles in the political system. The Presidency has been the principal national governing institution in the United States; its power has declined. The power of the media and of Congress has increased. Can their roles change? At this point the media are deeply committed to an opposition role. The critical question concerns the role of Congress. In the wake of a declining Presidency, can Congress organize itself to furnish the leadership to govern the country? During most of this century, the trends in Congress have been in the opposite direction. In recent years the increase in the power of Congress has outstripped an increase in its ability to govern. If the institutional balance between government and its opposition is to be redressed, the decline in Presidential power has to be reversed, and the ability of Congress to govern has to be increased.

Governing under distemper

The vigor of democracy in the United States in the 1960's thus contributed to a democratic distemper involving the expansion of governmental activity on the one hand, and the reduction of governmental authority on the other. This democratic distemper, in turn, had further important consequences for the functioning of the political system. The extent of these consequences, as of 1975, was still unclear, and dependent on the duration and the scope of the democratic surge.

The expansion of governmental activity produced budgetary deficits and a major expansion of total governmental debt from $336 billion in 1960 to $557 billion in 1971. These deficits contributed to inflationary tendencies in the economy. But at the same time that the expansion of governmental activity creates problems of financial solvency for government, the decline in governmental authority reduces still further the ability of government to deal effectively with these problems. The implementation of "hard" decisions imposing constraints on any major economic group is difficult in any democracy and particularly difficult in the United States, where the separation of powers provides economic interest groups with a variety of points of access to governmental decision-making. During the Korean War, for instance, governmental efforts at wage and price control failed miserably, as business and farm groups were able to riddle the legislation with loopholes in Congress, and labor was able to use its leverage with the executive branch to eviscerate wage controls. All this occurred despite the fact that there was a war on and the government was not lacking in authority. The decline in governmental authority in general and of the central leadership in particular during the early 1970's opens new opportunities for special interests to bend governmental behavior to their special purposes.

In the United States, as elsewhere in the industrialized world, domestic problems thus become intractible. The public develops expectations which it is impossible for government to meet. The activities—and expenditures—of government expand, and yet the success of government in achieving its goals seems dubious. In a democracy, however, political leaders in power need to score successes if they are going to stay in power. The natural result is to produce a gravitation to foreign policy, where successes—or seeming successes—are much more easily arranged than they are in domestic policy. Trips abroad, summit meetings, declarations and treaties, and rhetorical aggression all produce the appearance of activity and achievement. The weaker a political leader is at home, the more likely he is to be travelling abroad. The dynamics of this search for foreign policy achievements by democratic leaders who lack authority at home gives to dictatorships (whether Communist party states or oil sheikdoms)—which are free from such compulsions—a major advantage in the conduct of international relations.

The expansion of expenditures and the decrease in authority are also likely to encourage economic nationalism in democratic societies. Each country will have an interest in minimizing the export of some goods in order to keep prices down in its own society. At the same time, other interests are likely to demand protection against the import of foreign goods. In the United States, this has meant embargoes—as on the export of soybeans—on the one hand, and tariffs and quotas on the import of textiles, shoes, and comparable manufactured goods, on the other. A strong government will not necessarily follow more liberal and internationalist economic policies, but a weak government is almost certainly incapable of doing so.

Finally, a government which lacks authority and which is committed to substantial domestic programs will have little ability, short of a cataclysmic crisis, to impose on its people the sacrifices which may be necessary to deal with foreign policy problems and defense. In the early 1970's, as we have seen, spending for all significant foreign policy programs was far more unpopular than spending for any major domestic purpose. The United States government has given up the authority to draft its citizens into the armed forces, and is now committed to providing the monetary incentives to attract volunteers with a stationary or declining percentage of the GNP. At the present time, this would appear to pose no immediate deleterious consequences for national security. The question necessarily arises, however, whether, if a new threat to security should materialize in the future (as it inevitably will at some point), the government will possess the authority to command the resources, as well as the sacrifices, which are necessary to meet that threat.

For a quarter-century the United States was the hegemonic power in a system of world order. The manifestations of the democratic distemper, however, already have stimulated uncertainty among allies and could well stimulate adventurism among enemies. If American citizens don't trust their government, why should friendly foreigners? If American citizens challenge the authority of American government, why shouldn't unfriendly governments? The turning inward of American attention and the decline in authority of American governing institutions are closely related—as both cause and effect—to the relative downturn in American power and influence overseas. A decline in the governability of democracy at home means a decline in the influence of democracy abroad.

What's behind the democratic surge?

The immediate causes of the simultaneous expansion of governmental activity and the decline of governmental authority are to be found in the democratic surge of the 1960's. What, however, was responsible for this sharp increase in political consciousness, political participation, and commitment to egalitarian values?

The most specific, immediate, and, in a sense, "rational" causes of the democratic surge conceivably could be the specific policy problems confronting the United States government in the 1960's and 1970's, and its inability to deal effectively with those problems. Vietnam, race relations, Watergate, "stagflation": These quite naturally could lead to increased polarization over policy, higher levels of political participation (and protest), and reduced confidence in governmental institutions and leaders. In fact, these issues and the ways in which the government dealt with them did have some impact; the unraveling of Watergate was, for instance, followed by a significant decline in public confidence in the executive branch of government. More generally, however, a far from perfect fit exists between the perceived inability of the government to deal effectively with these policy problems and the various manifestations of the democratic surge.

The expansion of political participation was underway long before these problems came to a head in the mid-1960's, and the beginnings of the decline in trust in government go back before large-scale American involvement in Vietnam. Indeed, a closer look at the relationship between attitudes towards the Vietnam War and confidence in government suggests that the connection between the two may not be very significant. Opposition to U.S. involvement in Vietnam, for instance, became widespread among blacks in mid-1966; while among whites, opponents of the war did not outnumber supporters until early 1968. In terms of a variety of indices, however, white confidence and trust in government declined much further and more rapidly than black confidence and trust during the middle 1960's. In late 1967, for instance, whites were divided roughly 46 per cent in favor of the war and 44 per cent against, while blacks were split 29 per cent in favor and 57 per cent against. Yet in 1968, white opinion was divided 49.2 per cent to 40.5 per cent as to whether the government was run for the benefit of all or a "few big interests," while blacks thought that it was run for the benefit of all by a margin of 63.1 per cent to 28.6 per cent.[10] Black confidence in government plummeted only after the Nixon Administration came to power in 1969. While this evidence obviously is not conclusive, nonetheless it does suggest that the actual substantive character of governmental policies on the war—perhaps as well as on other matters—may be of less significance for the decrease in governmental authority than were other causes.

The democratic surge can also be explained in terms of the widespread demographic trends of the 1960's. Throughout the industrialized world during the 1960's, the younger age cohorts furnished many of the activists in the democratic and egalitarian challenges to established authority. In part, this revolt of the youth undoubtedly was the product of the global "baby boom" of the post-World War II years, which brought to the fore in the 1960's a generational bulge which overwhelmed colleges and universities. This was associated with the rise of distinctive new values which appeared first among college youth and then were diffused among youth generally. Prominent among these new values were what have been described by Daniel Yankelovich as "changes in relation to the authority of institutions such as the authority of law, the police, the government, the boss in the work situation." These changes were "in the direction of what sociologists call 'de-authorization,' i.e., a lessening of automatic obedience to, and respect for, established authority. . . ." The new disrespect for authority on the part of youth was part and parcel of broader changes in their attitudes and values concerning sexual morality, religion as a source of moral guidance, and traditional patriotism and allegiance to "my country right or wrong."[11]

As a result of this development, major differences over social

values and political attitudes emerged between generations. One
significant manifestation of the appearance of this generational gap
in the United States is the proportion of different age groups agreeing at different times in recent decades with the proposition: "Voting is the only way that people like me can have any say about
how the government runs things." In 1952, overwhelming majorities
of all age groups agreed with this statement, with only a one per
cent difference between the youngest age group (21-28), with 79
per cent approval, and the oldest age group (61 and over), with
80 per cent approval. By 1968, the proportion of every age group
supporting the statement had declined substantially. But of even
greater significance was the major gap of 25 per cent which had
opened up between the youngest age group (37 per cent approval)
and the oldest age group (62 per cent approval).[12] Whereas young
and old related almost identically to political participation in 1952,
they had very different attitudes toward it 16 years later.

Education and ideology

The democratic surge can also be explained as the first manifestation in the United States of the political impact of the social,
economic, and cultural trends connected with the emergence of a
post-industrial society. Rising levels of affluence and education
lead to changes in political attitudes and political behavior. Many
of the political and social values which are more likely to be found
among the young than among the elderly are also more likely to
be found among better-off, white-collar, suburban groups than
among the poorer, working-class, blue-collar groups in central and
industrial cities. The former groups, however, are growing in numbers and importance relative to the latter, and hence their political
attitudes and behavior patterns are likely to play an increasingly
dominant role in politics.[13]

The single most important status variable affecting political participation and attitudes is education. For several decades the level
of education in the United States has been rising rapidly. In 1940,
less than 40 per cent of the population was educated beyond ele-
mentary school; in 1972, 75 per cent of the population had been
either to high school (40 per cent) or to college (35 per cent). The
more educated a person is, the more likely he is to participate in
politics, to have a more consistent and more ideological outlook on
political issues, and to hold more "enlightened" or "liberal" or
"change-oriented" views on social, cultural, and foreign policy is-

sues. Consequently the democratic surge could be simply the reflection of a more highly educated populace.

This explanation, however, runs into difficulties when it is examined more closely. Verba and Nie, for instance, have shown that the actual rates of campaign activity which prevailed in the 1950's and 1960's ran far ahead of the rates which would have been projected simply as a result of changes in the educational composition of the population.[14] (In part, the explanation for this discrepancy stems from the tremendous increase in black political participation during these years.) In a similar vein, the assumption that increased attitude consistency can be explained primarily by higher levels of education also does not hold up. In fact, during the 1950's and 1960's major and roughly equal increases in attitude consistency occurred among both those who had gone to college and those who had not graduated from high school. In summarizing the data, Nie and Andersen state:

> The growth of attitude consistency within the mass public is clearly not the result of increases in the population's "ideological capacities" brought about by gains in educational attainment. . . . Those with the lowest educational attainment have experienced the largest increases in consistency on the core domestic issues; and little significant difference appears to be present between the two educational groups in comparison to the dramatic increases in consistency which both groups have experienced.

Instead, they argue, the increase in ideological thinking is primarily the result of the increased salience which citizens perceive politics to have for their own immediate concerns: "The political events of the last decade, and the crisis atmosphere which has attended them, have caused citizens to perceive politics as increasingly central to their lives."[15] Thus, the causes of increased attitude consistency, like the causes of higher political participation, are to be found in the changing political relationships, rather than in changes in individual background characteristics.

An excess of democracy

All this suggests that a full explanation of the democratic surge can be found neither in transitory events nor in secular social trends common to all industrial societies. The timing and nature of the surge in the United States especially need to be explained by distinctive dynamics of the American political process and, in particular, by the interaction between political ideas and institutional reality in the United States. Unlike Japanese society and most European societies, American society is characterized by a broad consensus favoring democratic, liberal, and egalitarian values. For much of the time, the commitment to these values is neither passionate nor intense. During periods of rapid social change, however, these democratic and egalitarian values of the American creed are reaffirmed. The intensity of belief during such "creedal passion periods" leads to the challenging of established authority and to major efforts to change governmental structure to accord more fully with those values. In this respect, as has already been remarked, the democratic surge of the 1960's shares many characteristics with the comparable egalitarian and reform movements of the Jacksonian and Progressive eras. Those "surges," like the contemporary one, also occurred during periods of realignment between party and governmental institutions on the one hand, and social forces on the other. The slogans, goals, values, and targets of all these movements are strikingly similar. Consequently, the implication of this analysis is that in due course the democratic surge and the resulting dual distemper in government will be moderated.

Al Smith once remarked, "The only cure for the evils of democracy is more democracy." Our analysis suggests that applying that cure at the present time could well be adding fuel to the fire. Instead, some of the problems of governance in the United States today stem from an "excess of democracy," in much the same sense in which David Donald used the term to refer to those consequences of the Jacksonian Revolution which helped to precipitate the Civil War. What is needed, instead, is a greater degree of moderation in democracy.

Democracy and moderation

In practice, this moderation has two major areas of application. First, democracy is only one way of constituting authority, and it is not necessarily a universally applicable one. In many situations, the claims of expertise, seniority, experience, and special talents may override the claims of democracy as a way of constituting authority. During the surge of the 1960's, however, the democratic principle was extended to many institutions where it can, in the long run, only frustrate the purposes of those institutions. A university where teaching appointments are subject to approval by students may be a more democratic university, but it is not likely to be a better university. In similar fashion, armies in which the commands of officers have been subject to veto by the collective wisdom of their subordinates have almost invariably come to disaster on the battlefield. The arenas where democratic procedures are appropriate are, in short, limited.

Second, the effective operation of a democratic political system usually requires some measure of apathy and non-involvement on the part of some individuals and groups. In the past, every democratic society has had a marginal population, of greater or lesser size, which has not actively participated in politics. In itself, this marginality on the part of some groups is inherently undemocratic, but it also has been one of the factors which has enabled democracy to function effectively. Marginal social groups, as in the case of the blacks, are now becoming full participants in the political system. Yet the danger of "overloading" the political system with demands which extend its functions and undermine its authority still remains. Less marginality on the part of some groups thus needs to be replaced by more self-restraint on the part of all groups.

The Greek philosophers argued that the best practical state—the "mixed regime"—would combine several different principles of government in its constitution. The Constitution of 1787 was drafted with this insight very much in mind. Over the years, however, the American political system has emerged as a distinctive case of extraordinarily democratic institutions joined to an exclusively democratic value system. Democracy, as a result, can very easily become a threat to itself in the United States. Political authority is never strong here, and it is peculiarly weak during a period of intense commitment to democratic and egalitarian ideals. In the United States, the strength of the democratic ideal poses a problem for the governability of democracy in a way which is not the case elsewhere.

The vulnerability of democratic government in the United States thus comes not primarily from external threats, though such threats are real, nor from internal subversion from the left or the right, although both possibilities could exist, but rather from the internal dynamics of democracy itself in a highly educated, mobilized, and participant society. "Democracy never lasts long," John Adams observed: "It soon wastes, exhausts, and murders itself. There never was a democracy yet that did not commit suicide." That suicide is more likely to be the product of overindulgence than of any other cause. A value which is normally good in itself is not necessarily optimized when it is maximized. We have come to recognize that there are potentially desirable limits to economic growth. There are also potentially desirable limits to the extension of political democracy. Democracy could have a longer life if it has a more balanced existence.

FOOTNOTES

[1] Governmental activity will be measured here primarily in terms of governmental expenditures. Such a measure, of course, does not do justice to many types of governmental activity, such as regulatory action or the establishment of minimum standards (e.g., for automotive safety or pollution levels or school desegregation), which have major impact on the economy and on society and yet do not cost very much in direct governmental spending.

[2] William Schneider, "Public Opinion: The Beginning of Ideology?" *Foreign Policy*, No. 17 (Winter 1974-75), pp. 88ff.

[3] Arthur H. Miller, "Political Issues and Trust in Government: 1964-1970," *American Political Science Review* 68 (September 1974), pp. 951ff.

[4] Gabriel A. Almond and Sidney Verba, *The Civic Culture* (Boston, Little Brown, 1965), pp. 64-68; Gallup Survey, New York *Times* (October 14, 1973), p. 45.

[5] Louis Harris and associates, *Confidence and Concern: Citizens View American Government*, Senate Committee on Government Operations, Subcommittee on Intergovernmental Relations, 93rd Congress, 1st Session (December 3, 1973).

[6] Richard W. Boyd, "Electoral Trends in Postwar Politics," in James David Barber, ed., *Choosing the President* (Englewood Cliffs, Prentice-Hall, 1974), p. 189.

[7] Gerald M. Pomper, "From Confusion to Clarity: Issues and American Voters, 1956-1968," *American Political Science Review* 66 (June 1972), pp. 415ff.; Miller, *op. cit.*, pp. 951ff.; Norman H. Nie and Kristi Anderson, "Mass Belief Systems Revisited," *Journal of Politics* 36 (August 1974), pp. 540-591; Schneider, *op. cit.*, pp. 98ff.

[8] Samuel H. Beer, "Government and Politics: An Imbalance," *The Center Magazine* 7 (March-April 1974), p. 15.

[9] Harris and associates, *op. cit.*, pp. 42-43, 299.

[10] See the University of Michigan Survey Research Center surveys of 1958, 1964, 1966, 1968, and 1970 on black and white "Attitudes Toward Government: Political Cynicism"; and John E. Mueller, *War, Presidents, and Public Opinion* (New York, John Wiley, 1973), pp. 140-48.

[11] Daniel Yankelovich, *Changing Youth Values in the '70's: A Study of American Youth* (New York, JDR 3rd Fund, 1974), p. 9.

[12] Anne Foner, "Age Stratification and Age Conflict in Political Life," *American Sociological Review* 39 (April 1974), p. 190.

[13] Samuel P. Huntington, "Postindustrial Politics: How Benign Will It Be?" *Comparative Politics*, Vol. 16, No. 2 (January 1974), pp. 177-82; Louis Harris, *The Anguish of Change* (New York, W. W. Norton, 1973), pp. 35-52, 272-73.

[14] Sidney Verba and Norman H. Nie, *Participation in America: Political Democracy and Social Equality* (New York, Harper and Row, 1972), pp. 251-52.

[15] Nie and Anderson, *op. cit.*, pp. 558-59.

2

The Declaration and the Constitution: liberty, democracy, and the Founders

MARTIN DIAMOND

In an address delivered in 1911, Henry Cabot Lodge, Sr., looked back wistfully to the joy with which, not long before, the country had celebrated two anniversaries: the centennial of the framing of the Constitution and, two years later, the centennial of its ratification. On both occasions, he remembered, great crowds thronged the streets, processions passed by amidst brilliant decorations and illuminations, and cannon and oratory thundered. What made the occasions so joyous, Lodge explained, was the conviction universally held by Americans of the original and continuing excellence of their Constitution: "Through all the rejoicings of those days . . . ran one unbroken strain of praise of the instrument and of gratitude to the men" whose wisdom had devised it. During those happy centennial celebrations, "every one agreed with Gladstone's famous declaration, that the Constitution of the United States was the greatest political instrument ever struck off on a single occasion by the minds of men."

But in words that speak to our own condition, Lodge then reflected with sorrow on the way all had changed so soon. The vast mass of the American people, he said, still believe in their Constitution, but now they "look and listen, bewildered and confused," as the air is "rent with harsh voices of criticism and attack." On

every side, they begin to hear opinion leaders, as we would call
them now, who raise a "discordant outcry" against that which the
people "have always reverenced and held in honor." Some "excellent persons" attack the Constitution out of an innocent but mis-
guided hope for human improvement by the easy means of constitutional alterations. Others attack the Constitution out of darker
motives and even greater folly: "Every one who is in distress, or
in debt, or discontented . . . every reformer of other people's mis-
deeds . . . every raw demagogue, every noisy agitator . . . all such
people now lift their hands to tear down or remake the Constitution." All such calling into question of the country's fundamental
political principles and institutions presented "a situation of utmost
gravity." "Beside the question of the maintenance or destruction of
the Constitution of the United States," Lodge solemnly concluded,
"all other questions of law and policies sink into utter insignificance."

Now, as we celebrate another great national anniversary, the Bicentennial of the American Revolution, the situation would seem
to be graver still. The "discordant outcry" of 1911, which Lodge
thought threatened our constitutional felicity then, has grown in the
course of 60 years from a few "harsh voices" out of tune with the
country's constitutional reverence into a majority of intellectual
voices, the conventional wisdom of those who give academic and
intellectual opinion to the nation. Moreover, not only has Lodge's
minority become our majority, but the then sketchy, somewhat
confused, and inconsistent theoretical foundations of his discordant
minority have been transformed into a formidable and comprehensive account of the founding of the republic. It is a disquieting
account which, quite apart from all other possible causes of political
distress, has itself the logical tendency to make impossible the kind
of constitutional contentment that so marked the nation's first cen-
tennials. And it is upon the basis of this disquieting account that
generations of American students have now received their instruction as to "what really happened" at the founding.

The nature of American democracy

Lodge's defense of the traditional understanding of the Constitution was in answer to Populist and Progressive demands for the
democratization of the Constitution, and thus of the American political order, through such means as the initiative, the referendum,
and the recall. During a generation of Populist and Progressive
attacks on the ugly plutocratic tendencies of the turn of the cen-

tury, these democratizing demands had brought steadily to the fore—as usually happens in American political debate—the question of the meaning and intention of the Constitution. The question took an historical form but—again, as is usually the case—had ultimately a philosophical and political significance: Was the Constitution originally democratic, only to have been perverted in later years, or was it simply undemocratic from the outset? The larger significance of the question, so formulated, is this: It assumes that *the* comprehensive political question is the extent to which things measure up to the requirements of the democratic form of government; democracy becomes the *summum bonum,* the complete good against which all else is judged.

It was around such an assumption concerning democracy, and the status of the Constitution relative to it, that everything ultimately proved to turn. But in 1911, as Lodge was speaking, his opponents were still divided in their answers. Most took the less grave view, namely, that the Constitution was originally satisfactorily democratic, had then been perverted, but that certain democratizing constitutional reforms would quite readily restore it to its original condition of excellence. Cast in such terms, this was still a debate among friends. Lodge and his opponents both agreed that the United States should have a democratic form of government and that the Constitution had established a good version of one. However, while Lodge was content that all was still satisfactorily democratic, his opponents were convinced otherwise. They proposed to graft some democratizing reforms on to the otherwise acceptable Constitution; he warned that these reforms were not really compatible with the rest of the Constitution, and chided them for their too simple faith in popular wisdom as directly and immediately applied to policy and legal questions. The issue between them came down to a question about the nature of democracy and how best to arrange or constitute it in America.

Now, this is a serious question but it is also a limited one, and it is an example of the archetypal American political question. That is to say, in contrast with the situation in many other countries, where the very nature of the regime is typically called into question, in the dispute between Lodge and his opponents there remained a consensus as to the fundamental type of regime the United States should have—namely, constitutional democracy—but serious, even bitter, conflict as to the particular constitutional arrangements. Each side could suspect the other of concealing more vehement aims under the common rhetoric of approval for constitutional democracy; and, if long exacerbated by concurrent social and economic discontents, the issues could of course escalate into profound conflict. But so long as both held to the agreement on the democratic character of the original Constitution, and claimed it as their standard, it remained a moderate debate within the boundaries of the American constitutional order.

Smith and the betrayal of the democratic promise

But even as Lodge was speaking, another, graver answer to the question about the democratic status of the Constitution was gathering force. Only four years earlier, in 1907, J. Allen Smith had published *The Spirit of American Government,* in which for the first time fully and passionately, a crucial turn was made in the Populist and Progressive argument. It was not true, Smith argued, that an originally democratic Constitution had been perverted to contemporary plutocratic purposes. Rather, democracy had never really gotten off the ground in America; the hopeful democratic promise of the Revolution had been deliberately betrayed from the very beginning by selfish "aristocratic" forces which had calculatedly framed the Constitution so as to prevent democratic majority rule and thus perpetuate privilege. Because democratic ideas had taken strong hold "among the masses," Smith said, the Founders were obliged to frame a government which conferred "at least the form of political power upon the people." But it was a sham conferral; they devised a "system of government which was just popular enough not to excite general opposition and which at the same time gave to the people as little as possible of the substance of political answer."

This was to deepen and darken the debate. The issue was no longer an intra-democratic dispute over the meaning and tendency of the Constitution, but a struggle between democracy and its foes, between democrats yearning to revive the old Revolutionary spirit and reactionaries entrenched behind the barriers of a Constitution, the spirit of which had always been anti-democratic. In principle, Smith's analysis invited those whom he succeeded in persuading to a far greater rejection of the entire constitutional system than anything contemplated by his predecessors. If the Constitution had indeed been the handiwork of a reactionary oligarchy, then all of its mechanisms and processes, its entire "spirit," must be understood as tending against the people and in favor of oligarchy. This was to create, in those who followed Smith's account of the founding, a frame of mind which was receptive to claims for radical political reconstruction.

And Smith's account—varied, elaborated, restated—became in time the dominant teaching regarding the founding among that part of American public opinion shaped by the intellectual and academic community. With that part of public opinion, the Constitution came to stand as if permanently on trial before a skeptical or hostile jury —and the more educated the jurors, the more skeptical or hostile they were likely to be. Nowadays we hear much of political alienation, of distrust in our fundamental institutions; while other factors are undoubtedly involved, this alienation is surely not unconnected with an account of the founding that nurtures a frame of mind prepared, to paraphrase Burke, to sniff radical constitutional defects "on every tainted breeze."

From Revolution to reaction?

Preeminent among those who taught a version of Smith's account of the founding are, of course, Charles A. Beard and Vernon L. Parrington. In his *An Economic Interpretation of the Constitution of the United States,* published in 1913, Beard developed an original version of the Smith view, emphasizing especially economic determinism, but in general manipulating data and conclusions in the manner that became characteristic of contemporary political science. Beard's influence was immense and—despite cogent rejoinders by many, especially E. S. Corwin—his argument came to be treated as having settled the fundamental question; for decades, government and history textbooks simply followed the Beard account. On another front, Vernon L. Parrington, whose *Main Currents in American Thought* (the first volume of which was published in 1927) was dedicated to J. Allen Smith, gave a general account of American literature which had its inspiration in Smith's account of the founding. Parrington built upon this base a general view of American life as the conflict between a crabbed old-world liberalism and a generous democratic outlook that derived from French romantic thought, with Hamilton and Jefferson starring as the rival champions. By the 1930's, in the worlds of both social science and literary criticism, Smith, Beard, and Parrington had achieved an amazing hegemony. Few who grew up in those years will have forgotten the sense of enlightenment, of emancipation, and of avant-garde intellectualism that came as one discovered these authors or their views.

Those views were encountered in the leading political science, history, and literary texts used in colleges and universities, in widely-read journals, and in scholarly and popular works generally. Among political science texts, examples are too numerous to need mentioning. One example among historians is the influential text of Samuel Eliot Morison and Henry Steele Commager, *The Growth of the American Republic*. The authors in their own way so fully accepted the dominant view that they presented a section of their chapter on the founding under the heading, "The Thermidorean Reaction." Thermidor! As if America under the Articles had had any resemblance to Robespierrean democracy, and as if the Constitution had anything like the anti-revolutionary character of French government after Thermidor. Under the impact, apparently, of the postwar revisionist attacks on Beard, this heading was silently dropped from later editions—but without, as I recall it, much change in the thrust of the argument. The chapter on the founding generation in Richard Hofstadter's *The American Political Tradition*, perhaps the most influential text among college students, also takes basically the same view. Hofstadter joins with an economic interpretation of the Framers a criticism of their outmoded Calvin-*cum*-Hobbes psychology, but the teaching is the same: democracy feared, democracy rejected; the Constitution as the expression of an outmoded undemocratic outlook still functioning as the barrier against which the forces of modern democracy must contend. A standard literary text—*Literary History of the United States*, by Spiller, Thorp, Johnson, Canby—deals with the founding political questions, understandably, somewhat more perfunctorily; but any lack of boldness in detail is amply made up for by the title of the chapter on the founding. It is "Revolution and Reaction," a favorite formulation of all those who take the Smith-Beard-Parrington position.

The overall character of this teaching constitutes a kind of drama in three acts, with a fourth act always trying to get itself written. As already indicated, Act I is the Revolution, with its brave Declaration of democratic equality and the hopeful democratic beginnings in the new states under the Articles of Confederation. Act II is the Constitution, reactionary and retrograde with respect to democracy, stifling the promising beginnings in the interest of economic privilege or in the name of a constricted, timorous, too jealous love of liberty. Act III is American history to date, in which frustrated democracy struggles against and gradually overcomes the confining Constitution; that is, by such means as constitutional amendment, legislation, and judicial interpretation, and through informal social and political developments, the country is thought to have become somewhat more democratic. However, since on this view the constricting Constitution is still there, and the struggle for democracy is not yet won, there is an Act IV always waiting to be written. The dramatic action here would consist in the final triumph of democracy. Democracy would no longer have to work its cramped way within the constraints of a Constitution always inimical to it, but would at last fully actualize itself and, by means of a new constitutional ordering, become the open, comprehensive, and fundamental law of the land.

So thoroughly has Lodge's "discordant outcry" become the conventional academic wisdom, that now it is the Constitution which seems to be the discordant element—an outmoded formal structure, either irrelevant to contemporary democratic needs or in conflict with them. No wonder there is a sense of faltering and uncertainty as we confront our Bicentennial. Abstracting from all other causes of concern, this understanding of the founding would be enough to darken the mood of the Bicentennial and of the reflections to which it gives rise.

Two views of the founding

The spirit in which we celebrate the Bicentennial will thus in considerable degree depend upon our understanding of the founding. If we continue to view it as fundamentally flawed—that is, as characterized by an improperly resolved question regarding the relationship of democracy and liberty—then we should unhesitatingly seize upon the Bicentennial as the occasion for new beginnings, as an opportunity to rally the nation toward a proper reconstituting of democracy. But if that view of the founding is false, in both history and political philosophy, then we ought, equally unhesitatingly, to make the Bicentennial the occasion for renewed appreciation of our fundamental institutions and rededication to their perpetuation.

But which view of the founding is the right one? The now dominant view or the traditional one it supplanted? The question may be conveniently discussed in terms of our two fundamental documents, the Declaration and the Constitution, and the relationship between liberty and democracy as expressed in them. These are the two great charters of our national existence, representing the beginning of our founding and its consummation; in them are incarnated the two principles—liberty and democracy—upon the basis of which our political order was established, and upon the understanding of which in each generation our political life in some important way depends. Inevitably, then, the dispute about the founding has in a sense always revolved about the understanding of the Declaration and the Constitution and how they stand toward each other.

Now, all parties to the dispute agree that one of these documents stands somehow fundamentally for liberty, and the other one for democracy. But which is which? Interestingly and conveniently, in the two conflicting views of the founding the status of the two documents is exactly reversed. In the now dominant view, the Declaration is understood as a democratic manifesto, or at least as positing principles that logically point to the democratic form of government; and the Constitution is understood as having turned against the democracy of the Declaration in the name of liberty, but in actuality of privilege masquerading as liberty. But in the older view, the one held by the leading men who wrote the two documents—and the correct view, as I shall argue—the matter was understood just the other way round. The Declaration is understood as holding forth only the self-evident truths regarding liberty and, thus, as doing no more than declaring liberty the only legitimate end of all government, whatever its form; and it is the Constitution which is viewed as having opted for democracy, as embodying the bold and unprecedented decision to achieve, in so large a country as this, a free society under the democratic form of government.

Now there is something startling to contemporary Americans, something unpersuasive on its face, in this old-fashioned view of the relation between the Declaration and the Constitution, of libery and democracy. How can it really be, one might ask, that the glowing, resounding, revolutionary Declaration was limited to the principle of liberty while the sober—one might say, phlegmatic—Constitution represented the bold establishment of democratic government? But the old-fashioned view is startling only to those who have made democracy, understood as the achievement of human equality in every respect, *the* comprehensive political good, superior to liberty or comfortably comprehending it. Eighteenth-century Americans had not yet become so complacent about democracy. Indeed, the founding generation would have balked at the language used above in speaking of the Declaration as *limited* to liberty, as *only* making liberty the end of government. They could not have accepted the implication that democracy was something higher than liberty and beyond it in worth and dignity. On the contrary, *for the founding generation it was liberty that was the comprehensive good, the end against which political things had to be measured; and democracy was only a form of government which, like any other form of government, had to prove itself adequately instrumental to the securing of liberty.*

The horizon of liberty

To understand the founding, then, we must understand anew the horizon or perspective of liberty within which the Declaration and the Constitution were conceived, and the status of democracy within that horizon. A statement of Hamilton from *The Federalist*, which it should be remembered Jefferson unreservedly regarded as the authoritative commentary on the Constitution, will help to renew our understanding. In *The Federalist* Number 9, Hamilton is defending the idea of popular republican government (by which, he says elsewhere, he meant "representative democracy") from its critics. He acknowledges, first, the force of their argument; "it is not to be denied," he says, that the history of all earlier popular republics is so wretched, so disfigured by disorders, as to make plausible the critics' contempt of popular government. Indeed, if there were not some new, improved way to devise a popular republic, then "the enlightened friends to liberty would have been obliged to abandon the cause of that species of government as indefensible." Hamilton of course proceeds to argue that the Constitution will establish just such a novel and defensible system. But the crucial point for us here is the priority of liberty as the end of government, the merely instrumental status of all forms of government, and the peculiarly questionable status of the popular form—democracy—up to the time of the American founding. I cannot conceive of a single author of the Declaration or a leading Framer of the Constitution who would have disagreed with Hamilton's formulation. They all saw the relationship of the Declaration and the Constitution in exactly the same light.

Now this liberty, the preciousness and precariousness of which are now obscured by a complacent contemporary egalitarianism, seemed then a very grand and novel idea to those who drafted our two national documents. They regarded liberty as a modern idea, as the extraordinary achievement of 17th- and 18th-century political thought. With that achievement in mind, George Washington, for

example, said that Americans lived in "an Epocha when the rights
of mankind were better understood and more clearly defined, than
at any former period." This understanding and clarity were all part
of the new "science of politics" to which Hamilton referred. It
rested, as Leo Strauss has persuasively explained, upon a crucial
turn in political thought made by the great philosophic predecessors
of the Americans—a turn toward practicable liberty rather than
utopian virtue as the end at which governments ought to aim. For
thinkers like Locke and Montesquieu it thus became the task of
government to provide the framework for a safe and free society in
which all equally could enjoy their "unalienable rights ... [to] life,
liberty, and the pursuit of happiness." Such prescriptions for political liberty, the American Founders thought, gave hope at last for
the "relief of man's estate," and it was with such prescriptions in
mind that they drafted the Declaration and the Constitution.

Equality as equal political liberty

Once this is understood, one can turn directly to the text of the
Declaration and hope to read it as it was intended. Here is the crucial passage:

> We hold these Truths to be self-evident, that all Men are created
> equal, that they are endowed by their Creator with certain unalien-
> able Rights, that among these are Life, Liberty and the Pursuit of
> Happiness—That to secure these Rights, Governments are instituted
> among Men, deriving their just Powers from the Consent of the Gov-
> erned, that whenever any Form of Government becomes destructive
> of these Ends, it is the Right of the People to alter or to abolish it,
> and to institute new Government, laying its Foundations on such prin-
> ciples, and organizing its Powers in such Form, as to them shall seem
> most likely to effect their Safety and Happiness.

Two of the key phrases—"created equal" and "consent of the
governed"—have been particularly misunderstood because they have
been wrenched out of context. Written within the horizon of liberty
of the founding generation, they have been understood instead
within the horizon of egalitarian democracy to which later generations have tended. The Declaration does not mean by "equal" anything at all like the general human equality which so many now
make their political standard. Jefferson's original draft of the Declaration is especially illuminating in this respect. All men, Jefferson
first wrote, "are created equal and independent" and from that
"equal creation they derive rights inherent and inalienable." The
word "independent" is especially instructive; it refers to the condi-

tion of men in the state of nature. In fact, this famous passage of the Declaration, in both its original and final formulations, must be understood as dealing entirely with the question of the state of nature and with the movement from it into political society.

The social contract theory upon which the Declaration is based teaches not equality as such but equal political liberty. The reasoning of the Declaration is as follows. Each man is equally born into the state of nature in a condition of absolute *independence* of every other man. That equal independence of each from all, as John Locke put it, forms a "Title to perfect Freedom" for every man. It is this equal perfect freedom, which men leave behind them when they quit the state of nature, from which they derive their equal "unalienable rights" in civil society. The equality of the Declaration, then, consists entirely in the equal entitlement of all to the rights which comprise political liberty, and nothing more. Thus Lincoln wisely interpreted the Declaration: "The authors of that notable instrument . . . did not intend to declare all men equal in all respects. They did not mean to say all were equal in color, size, intellect, moral developments, or social capacity. They defined with tolerable distinctness, in what respects they did consider all men created equal—equal in 'certain inalienable rights, among which are life, liberty, and the pursuit of happiness.'"

Now "to secure these rights," men quit the insecure state of nature and "Governments are instituted among men, deriving their just powers from the consent of the governed." Here we have the unambiguous meaning of the other phrase in the Declaration that is now so typically misunderstood. It has been transformed to mean rule by the consent of majorities, that is, consent according to the procedures of the democratic form of government. *But the Declaration does not say that consent is the means by which the government is to operate; it says that consent is necessary only to institute or establish the government.* It does not prescribe that the people establish a democratic form of government which *operates* by means of their consent. Indeed, the Declaration says that they may organize government on "such principles" as they choose, and that they may choose "any form of government" they deem appropriate to secure their rights. (In this, the Declaration was again simply following Locke, who taught that when men consent "to joyn into and make one Society," they "might set up what form of Government they thought fit.") And by "any form of government," the Declaration emphatically includes—as any literate 18th-century reader would have understood—not only the democratic form of

government, but also a mixed form, and the aristocratic and monarchic forms as well. That is why, for example, the Declaration has
to submit facts to a "candid world" to prove the British king guilty
of a "long train of abuses." Tom Paine, by way of contrast, could
dispose of King George more simply. Paine deemed George III unfit
to rule simply because he was a *king* and kingly rule was illegitimate as such. The fact that George was a "Royal Brute" was only
frosting on the cake; for Paine his being royal was sufficient warrant
for deposing him. But the Declaration, on the contrary, is obliged
to prove that George was indeed a brute. That is, the Declaration
holds George III "unfit to be the ruler of a free people" not because he was a king, but because he was a *tyrannical* king. Had the
British monarchy continued to secure to the colonists their rights,
as it had prior to the long train of abuses, the colonists would not
have been entitled to rebel. It was only the fact, according to the
Declaration, that George had become a tyrannical king that supplied the warrant for revolution. Indeed, most of the signers of the
Declaration probably would have cheerfully agreed that the English "mixed monarchy" had been, and perhaps still was for the
English themselves, the best and freest government in the history
of mankind.

Thus the Declaration, accurately speaking, is neutral on the
question of forms of government; any form is legitimate, provided
it secures equal freedom and is instituted by popular consent. But
as to how to secure that freedom the Declaration, in its famous
passage on the principles of government, is silent.

The "inconveniences of democracy"

The Framers of the Constitution were not lacking in guidance
from that same new science of politics which had taught them to
make liberty the end of government. Especially in Montesquieu's
The Spirit of the Laws, there was guidance for the construction of,
and statesmanship within, the various forms of government. Montesquieu understood that every form of government has its own
peculiar excellences, and also its own peculiar tendency to corruption or degeneration. In this, Montesquieu was simply in agreement
with the great tradition of political thought. For example, in Aristotle's fundamental typology of regimes, there are six basic forms—
three that are "true" forms, namely, monarchy, aristocracy, and
"polity," and their three perversions, that is, the forms into which
each tends to degenerate, namely, tyranny, oligarchy, and "demo-

cracy." On the basis of this typology, Aristotle teaches how to arrange or constitute the regime, whatever type it is, so as to help it be its distinctive best self in its circumstances, and so as to guard against its distinctive degenerative tendency. Montesquieu likewise has a typology of regimes (or forms of government, more precisely), and an analysis of their peculiar excellences and degenerative tendencies. In this regard, then, despite the vast differences between them, Aristotle and Montesquieu are at one.

Their common reasoning is as follows. Every regime or form of government has to have a "ruling element," some part of the whole —whether it be the monarch, the few, or the many—which has to have the final political power. But that power is always susceptible to perversion or abuse. In each political order, the slide to perversion or abuse follows the path of weakness peculiar to that order, and prudent founders and statesmen take pains to guard against the slide.[1]

If we understand that the American Founders, like all sensible men before them, regarded *every* form of government as problematic, in the sense of having a peculiar liability to corruption, and accepted the necessity to cope with the problematics peculiar to their *own* form of government, we will be able to emancipate ourselves from a particularly misleading and persuasive error made by those who follow the now dominant account of the founding. In this account, much is made argumentatively and rhetorically of the fact that the Framers frequently delivered themselves of very sharp criticism of the defects and dangers of democracy. Perhaps the favorite exemplary quotation is that of Edmund Randolph who, at the Federal Convention, traced the evils of the day to their origin in the "turbulence and follies of democracy." Generations of students have shared with their instructors the titillating satisfaction of shocked disdain for the crudely anti-democratic bias of the Foun-

---

[1] To emphasize their common reasoning in this matter is not, however, to minimize the vast difference between Aristotle, the great exponent of ancient political science, and Montesquieu, of the modern. Aristotle has in mind the enhancement in each regime, to the extent possible, of a high conception of virtue and justice; Montesquieu has in mind the security of individual liberty. In accordance with this difference, Aristotle and Montesquieu categorize the various kinds of regimes differently, diagnose differently what their degenerative tendencies are, and prescribe differently because of the different things each is trying to achieve or protect in the various regimes. This difference between the ancient and modern views is illuminated in an observation of Leo Strauss. From the point of view of modern thought, he says, "what you need is not so much formation of character and moral appeal [which ancient thought required] as the right kind of institutions, institutions with teeth in them." The American Founders followed Montesquieu in their reliance on institutions, and not the ancients regarding the necessity of character formation.

ders, as revealed in that remark and similar utterances. The inference drawn from such criticisms of the peculiar defects of democracy has been that the Founders naturally rejected what they regarded as defective. But we may now understand the matter in a different light. *Of course,* the Founders criticized the defects and dangers of democracy and did not waste much breath on the defects and dangers of the other forms of government. For a very good reason. They were not founding any other kind of government; they were establishing a democratic form, and it was the dangers peculiar to it against which all their efforts had to be bent.

An especially clear example, as always, is James Madison's discussion of the separation of powers as it was structured in the proposed Constitution. The separation of powers, he explains in *The Federalist* Number 48, should be differently arranged in many vital details, according to the form of government with which one is dealing. Where there is an hereditary monarch, for example, or in a small direct democracy, the danger of corruption and abuse is different from what it will be in the "representative republic" that is being established by the Constitution. And in accordance with the difference in the danger, the allocations and arrangements of the separated powers likewise ought to be different. Madison spends very little time, however, in discussing the provisions that should be made against dangers elsewhere; rather, he devotes himself to the problems of the form of government at hand—namely, the democratic republic to be established under the Constitution. Sufficient unto a Founder are the evils of the form he is founding.

Properly understood, then, the extent and intensity of the founding generation's concern for the defects and dangers of the democratic form, far from indicating their rejection of democracy, is proof of their acceptance of it and of their determination—copiously expressed, if only one will listen to them—to cope with it. Thus Madison coolly analyzed the "inconveniences of democracy," but only in order to deal with them in a manner "consistent with the democratic form of Government." Similarly, in *The Federalist* Number 9, Hamilton claims that the Founders are employing the newly discovered and improved means "by which the excellencies of republican government may be retained and its imperfections lessened or avoided." And above all, there is Madison, whose Number 10 of *The Federalist* is rightly regarded as the most important original political writing by an American. In it Madison presents his solution to the old problem of majority faction (the tyranny of the majority, as we call it) which had earned for popular governments the just "opprobrium" under which they had hitherto always labored. Madison proudly claims that under the Constitution that form of government will be rescued and can at last "be recommended to the esteem and adoption of mankind."

How utterly the intention and achievement of the Founders are hidden from us by an account of the founding which denies them their democratic *bona fides!* That intention consisted above all in trying to establish a system faithful to the "spirit and form of popular government" in which individual liberty would be made secure. It was the "honorable determination," Madison wrote in *The Federalist* Number 39, of every American "votary of freedom to rest all our political experiments on the capacity of mankind for self-government." Here we have the very heart of the matter: As votaries of freedom, *individual liberty was to the Founders the comprehensive, unproblematic good; and they were determined to secure that good by an experiment in democracy.*

The political science of liberty

Now wherein lay the difficulty and the danger? We may turn to Montesquieu for guidance to the Founders' understanding of their task. Accepting Locke's perspective on liberty, Montesquieu was the great teacher of the political science of liberty, of the means by which it could be made secure in various forms of government, or, failing that, then at least means by which constraints upon liberty could be moderated. In Book XI of *The Spirit of the Laws* Montesquieu states the problem. No form of government is free "by its nature"; that is, none has liberty built securely into its very form. This is because, as observed above, every form of government consists in giving power to some ruler or body of rulers. And "eternal experience" teaches that, if this power is unrestrained, it will inevitably come to be abused. But how to restrain that necessary power? It is in the very "disposition of things" that power can only be restrained by another power. This restraint, this prevention of unrestrained political power, is for Montesquieu the very definition of political liberty. "The political liberty of the citizen is a tranquility of mind arising from the opinion each has of his safety." When will the citizen be warranted in holding such an opinion? The answer is: when, within the bounds of the possible, his government has been so constituted that no other citizen or body of citizens is unrestrainedly able to employ the powers of government to oppress him.

It is at precisely this point that we can see the modernity of Montesquieu. Montesquieu does not seek to restraint rulers (except, interestingly, in despotisms) by means of the teachings of religion, or by such instruction in virtue as will moderate them or make them friends to liberty. Nor does he rely primarily on the traditional prescription of the "mixed regime," that means of arranging government so that both Lords and Commons, both patricians and plebians, both oligarchs and demos, have to concur in the public measures. Montesquieu turns instead toward a purely institutional means for securing liberty—above all, the separation of powers. His political science is a political science of institutions (and also, it can only be mentioned here, of such a "constitutional" arrangement of the interests and passions as will be conducive of the same mild results sought through institutional means).

In the 19th century, there were many who mocked Montesquieu for his fear of political power and for his cautious institutional strategies. Such men had a confidence in power on the basis of "a conception of human nature," as Parrington described it in another context, "as potentially excellent and capable of indefinite development." But let those now mock who read the 20th century as warranting credence in such a conception of human nature, as entitling men to adventures in unrestrained power. The American Founders, at any rate, preferred to follow men like Montesquieu and the teachings of modern liberty, and chose to find democratic means under the Constitution to secure that liberty. Their great merit consisted in taking the political science of liberty, as that is expressed in the Declaration, and elaborating it into its first full application to the democratic form of government under the Constitution. That was their intention. This has been shown. That it was also their achievement would require an argument beyond the scope of the present essay. But this can be said: It is an intention and achievement which anyone raised under the present scholarly dispensation is incapable of understanding and appreciating.

The American posture toward democracy

From the relationship of the Declaration and the Constitution, as it was understood by the founding generation, there emerges what can be called the American posture toward democracy. Henry Cabot Lodge, with whose post-Centennial reflections this essay began, understood that posture well because he understood the founding in the traditional way which I have attempted to restore:

"The makers of the Constitution . . . knew that what they were establishing was a democracy." But "the vital question was how should this be done." With the history of democracy's past failures in mind, they were determined to "so arrange the government that it should be safe as well as strong. . . . They did not try to set any barrier in the way of the popular will, but they sought to put effective obstacles in the path to sudden action which was impelled by popular passion, or popular whim. . . ." This was the issue as it confronted Lodge: Should the American posture toward democracy remain that of self-doubt and self-restraint, of reliance on representative institutions, on the separation of powers, and all the other self-imposed moderating devices of the Constitution? Or should the political thrust of society be in the direction of government by a more immediate and direct popular will?

For us the issue has become very much graver. With us now, it is not just a question of imprudent democratizing reforms, but a vast inflation of the idea of equality, a conversion of the idea of equal political liberty into an ideology of equality. The underlying complaint against the American political order is no longer a matter of mere reforms, or even of wholesale constitutional revision, although there is always a kind of itch in that direction, but rather a critique of the entire regime in the name of a demand for equality in every aspect of human life. It is a demand which consists in a kind of absolutization of a single principle, the principle of equality, and at the same time an absolutization of the democratic form of government understood as the vehicle for that complete equality. This is a different posture toward democracy indeed than that embodied in the American founding.

The deepest political question Americans can ponder during their Bicentennial celebration is precisely this rivalry of democratic postures or outlooks, the original one of the founding and the newer one based on egalitarianism. The pondering of this question involves considerations far beyond the scope of this essay. But this much can be said: It is a rivalry that will not be fairly contested, a question that will not be wisely pondered, so long as the American political order is understood in the light of the account of the founding that I have been criticizing. That account prejudges the question; it denies democratic credentials to the traditional American posture toward democracy and thereby tilts the scales in favor of egalitarian claims against the present constitutional order. The Bicentennial is a good occasion for the restoration of those credentials.

3

The past and future Presidency

AARON WILDAVSKY

IN the third volume of *The American Commonwealth*, Lord Bryce wrote, "Perhaps no form of Government needs great leaders so much as democracy." Why, then, is it so difficult to find them? The faults of leadership are the everyday staple of conversation. All of us have become aware of what Bryce had in mind in his chapter on "True Faults of American Democracy," when he alluded to "a certain commonness of mind and tone, a want of dignity and elevation in and about the conduct of public affairs, an insensibility to the nobler aspects and finer responsibilities of national life." If leaders have let us down, they have been helped, as Bryce foresaw, by the cynical "apathy among the luxurious classes and fastidious minds, who find themselves of no more account than the ordinary voter, and are disgusted by the superficial vulgarities of public life." But Bryce did not confuse condemnation with criticism. He thought that "the problem of conducting a stable executive in a democratic country is indeed so immensely difficult that anything short of failure deserves to be called a success. . . ." Explaining "Why Great Men Are Not Chosen," in the first volume of his classic, Bryce located the defect not only in party politics but in popular passions: "The ordinary American voter does not object to mediocrity."

Ultimately, Bryce was convinced, "republics Live by Virtue"—with a capital "V," meaning "the maintenance of a high level of public spirit and justice among the citizens." Note: "among the citizens," not merely among public officials. For how could leaders rise so far above the led; or, stemming from the people, be so superior to them; or, held accountable, stray so far from popular will? Surely it would be surprising if the vices of politicians stemmed from the virtues of the people. What the people do to their leaders must be at least as important as what the leaders do to them. There are, after all, so many of us and so few of them. Separating the Presidency from the people—as if a President owed everything to them and they nothing to him—makes as much sense as removing the people from the government it has instituted.

If the reciprocal relations between political leadership and social expectations could be resolved by exhortation, the problem would long ago have ceased to be serious. If expectations are not being met, it is leaders who are not meeting them, and either lower expectations or higher caliber leadership is required. But both may be out of kilter. Should one or two leaders fail, that may well be their fault. When all fail (Kennedy, Johnson, Nixon, and now Ford), when, moreover, all known replacements are expected to fail, the difficulty is not individual but systemic: It is not the action of one side or the reaction of another but their mutual relationships that are flawed. That the people may reject their Presidents is obvious; that Presidents might flee from their people is less so. Once Presidents discover the embrace of the people is deadly, they may well seek to escape from it. Presidents who tried to exercise powers they did not have might then be replaced by Presidents unwilling to exercise the powers they do have. The future of the Presidency will be determined not by the Presidency alone but by how Presidents behave in response to the environment "We-the-People" create for them. Presidents, we shall learn, can retreat as well as advance.

Presidential popularity

In the future, Presidents will be more important but less popular than they are today. The Presidency will be more powerful vis-à-vis institutional competitors but less able to satisfy citizen preferences than it is now. Unwilling to play a losing hand, future Presidents will try to change the rules of the game. If they cannot get support from the people, they can increase the distance between themselves and their predators.

The importance of Presidents is a function of the scope of government; the more it does, the more important they become. No one believes that the federal government of the United States is about to do less—to abandon activities in which it is now engaged, to refuse involvement in new ones, to go below instead of above the per cent of the Gross National Product it now consumes. Hence Presidents must, on the average, be more important than they used to be. Even if one assumes the worst—a weak President opposed by strong Congressional majorities—the President's support will make it easier for proposals to receive favorable consideration and his opposition will make it less likely that legislation will be considered at all or be passed over his veto. Limiting Presidential importance would require the one thing no one expects—limiting what government does.

Presidents remain preeminent in foreign and defense policy. Formal authority is theirs; informal authority, the expectations as to who will do what, is almost entirely in their domain. There does not have to be a discernible foreign policy but, if there is, Presidents are the people who are expected to make it. The exceptions—the Jackson Amendment on Soviet Jews, restrictions on aid to Cambodia, Vietnam, and Turkey—prove the rule. They show that Presidents are not all-powerful in foreign policy; they do not get all they want when they want it. But these incidents are just what they seem —minor. The Turks would not be behaving differently in Cyprus if they continued to receive American aid, and détente, if there ever was such a thing, has evidently not been disturbed by Russian repudiation of their undertaking on the emigration of Soviet Jews. Reluctance to provide funds for Cambodia and Vietnam was, at most, a matter of deciding when—under the present Republican or a future Democratic President—rather than whether these government would fall. Congress may anger foreign governments and dismay Secretaries of State; it may point to itself as evidence of national disagreement; but it will not succeed in making foreign policy. The main recommendation of the Congressional delegation to Cambodia, after all, was to send the Secretary of State there post haste!

I risk belaboring the obvious because, in the backlash of Watergate, it has become all too easy to imagine a weakening of the Presidency. Not so. Does anyone imagine fewer groups will be interested in influencing a President's position in their own behalf or that his actions will matter less to people in the future? The question answers itself. The weakening of the Presidency is about as likely as the withering away of the state.

To be important, however, is not necessarily to be popular. Let us conceive of Presidential popularity as a vector of two forces: long-term dispositions to support or oppose the institution and short-run tendencies to approve or disapprove what the occupant of the office is doing. Either way, I believe, the secular trend in popularity will be down.

By far the most significant determination of Presidential popularity is the party identification of the population. Since the proportion of people identifying with the major parties has shown a precipitous decline, future Presidents are bound to start out with a smaller base comprised of less committed supporters. This tendency will be reinforced by a relative decline of the groups—the less educated and the religious fundamentalist—who have been most disposed to give unwavering support regardless of what a President does or fails to do. Education may not make people wise, but it does make them critical. As the number of critical people in the country increases, criticism of Presidents will naturally increase. Thus future Presidents will have to work harder than have past Presidents to keep the same popularity status, so to speak. To offset long-run decline they will count on good news in the short run. But the news is bound to be bad.

Public policy and political responsibility

There is no consensus on foreign policy; the lessons learned from the past have not proved helpful. The 1930's apparently taught the United States to intervene everywhere, and the 1960's to intervene nowhere. Neither lesson is supportable. Under the spell of Vietnam, the instinctive reaction to foreign policy questions is "no." Foreign policy requires faith: Evils must be avoided before their bloody consequences manifest themselves to the doubtful. But there is no faith. As in the parable of the Doubting Thomas, Congressmen would not believe the President concerning the situation in Cambodia until they actually went there, plunged their hands into the wounds of the people, and saw they were red. For Presidents the adaptive response will be inaction until it is too late, after which there is no point in doing anything.

Reaction to the oil embargo and the manyfold increase in oil prices is a portent of things to come. When it came in the midst of an Egyptian and Syrian invasion of Israel, the United States did not react at all. Why were the American people not told of the inevitable consequences of the oil price increase, from mass starvation in poor countries and financial havoc among allies to inflation at home? Because then the President would have had to do something about it.

Future Presidents will allow foreign events to speak for themselves after the fact so they don't have to speak to them beforehand. They may reluctantly give in to popular demands for strong action but they will not act in anticipation. Followership, not leadership, probably will best describe future Presidential foreign policy.

There is not today, nor is there likely to be in the near future, a stable constituency in support of social reform. The 1930's through the 1950's were easy to understand. The "haves" did not like to pay, and the "have nots" preferred the benefits they received to the alternative. By the late 1960's, however, the poorer beneficiaries had learned from their leaders that they did not benefit, which led the richer providers to add ingratitude to their list of complaints. Then the extreme passion for equality, against which Tocqueville warned, asserted itself to label anything the government was able to accomplish unworthy of achievement. Too little, too late; or too much, too soon, the result was always the same: a feeling of failure. For as long as directly contradictory demands are made on public policy, governments (and hence Presidents) will be unable to get credit for what they do.

Public housing is a good example of how to make an evident failure out of an apparent success. The nation started with a low-cost shelter program for the stable working poor. Public policy concentrated the available resources into housing projects, so some people could be helped; and resident managers used their discretion to screen out "undesirables," so tenants could live in peace. It was not necessarily the best of worlds for all, but it was better for some. But all that changed in the 1960's. Screening was condemned as racism and worse: Didn't justice require that the worst-off be given preference? Not long after, public housing was attacked as a failure for all the crime it attracted. The dynamiting of buildings in St. Louis' Pruitt-Igoe Project symbolized, unfairly but persuasively, the blowup of hopes for public housing.

More medical care for the poor and aged is incompatible with easier access and lower costs for the whole population. By now the poor see doctors about as often as anybody else. This might be considered an accomplishment of Medicare and Medicaid— except that no one wants to accept responsibility for the consequences. Because access increases faster than facilities, the medical system gets crowded; because doctor and patient are motivated to resolve their uncertainties about treatment by using the insurance and subsidies at their disposal, the system gets expensive; because medical care is only moderately related to health, morbidity does not decline and mortality does not decrease in proportion to expenditure. Hence we hear that the system is in crisis because it is overcrowded, too expensive, and doesn't do much to improve aggregate health rates.

Examples of incompatibilities could come from almost any area of public policy—training the hard-core unemployed at low cost or decreasing dependency by increasing payments to people on welfare—but I will content myself with one less obvious example: party reform. The evils besetting our major political parties are supposed to be excessive influence of money and insufficient power of participation. But participation requires more meetings, conferences, primaries; in a word, more money. Told that money is the root of all evil, on the one hand, and required to dig deeper for it to promote participation, on the other, the position of our parties must be as perilous as that of Presidents, who are urged simultaneously to limit their powers and to lead the people.

Where will support for Presidential power come from? The only time a constituency appears is when there is a threat of curbing or eliminating existing programs. Then one discovers there are really recipients of medical care, aid to dependent children, food stamps, and the whole panoply of subsidies, transfer payments, and tax expenditures. But when they are safe, one could never guess from the torrent of abuse (money matters more than men, forms triumph over functions) that they had something worthwhile to lose. But they do and they will.

The mega-increase in the cost of energy means a decline in our standard of living. We pay more and get less. There will be fewer resources available to support social programs. Race relations may worsen as the poor (and black) do worse. Conservationists and producers will disagree more over strip-mining, oil shale, atomic power plants, and the like. Contradictory demands on government —produce more energy with less damage to the environment at lower cost—will increase.

It will be difficult to reduce defense expenditures because allies and dependents will be poorer and weaker than they were. It will be hard to avoid the threat of force because international events will become more threatening, and it will be difficult to use force because the nation will be torn between recent memories of Viet-

nam ("No more foreign adventures!") and older recollections of
the Second World War ("Intervene before it is too late!"). As foreign news becomes less favorable, government will seek to apply
more pressure at home. Is a tariff on imported oil a foreign or a
domestic policy? Obviously, it is both. Yet there is no reason to
believe that people will become less attached to their lifestyles or
less interested in benefits from expensive spending programs. Demands on government will increase as the willingness of citizens
to pay for them decreases. Reduction and redirection of consumption will become inevitable. The lot of a President will not be a
happy one.

In view of these circumstances, the barest extrapolation from
current events, it hardly seems likely that the nation will have to
worry about a too powerful Presidency, a legislative dictatorship
(as President Ford claimed in over-zealous campaign rhetoric), judicial tyranny, or any of the other scare slogans of the day. There
will be enough blame for everyone. The complaint will be that
our political institutions are too weak in comparison to their responsibilities, not that one is too strong in relation to another.

Presidential power

If it is a question of who gets the blame, a President will still
be first in war, in peace, and in the hearts of his countrymen, for
he will remain the single most visible and most accountable political actor. If it is a question of whose views will prevail on the
largest number of important issues over the longest period of time,
Presidents will still beat out their competitors.

In order to be a consistent competitor Congress would have to
speak with a single voice. The days are gone when House Speaker
"Uncle" Joe Cannon held so much internal power that Presidents
Theodore Roosevelt and William Howard Taft had to deal with
him as if he were a foreign potentate. I refer not merely to independent Congressional election through a decentralized party system in which no person owes election to another (though it is
worth reminding ourselves of this fact of life) but also to the continuing dispersal of power. The tendency of every procedural reform (save one) is to disperse substantive power. The more Congressmen are forced to work out in the open, the less they are able
to concert with one another in private. The more difficult it is to
compromise their differences, the less they are able to oppose the
Chief Executive. The decline of seniority may permit talent to be

substituted for age; it also guarantees that Congress will provide less attractive careers for people with whom Presidents have had to deal, not once but indefinitely. Participatory democracy may have many virtues but institutional cohesion is not likely to be one of them. Enhancing the ability of individual legislators to express themselves is not equivalent to uniting them in behalf of a common program.

The exception is the budget reform. If successful, it would, by relating revenue to expenditure, enable Congress to maintain its power of the purse. But the prognosis is problematic. The impetus for change came from the excessive and extreme use of impoundment by President Nixon. Whether this was only a temporary abuse, or represents adaptation to a situation in which Congressmen like to get credit for spending but blame Presidents for taxation, remains to be seen.

If the new House and Senate budget committees attempt to act like cabinets on the British model, which enforce their preferences on the legislature, they will fail. Cabinets are committees that tolerate no rivals and Congress is composed of rivals. If the budget committees permit too many deviations, or are overruled too often, it will become clear that expenditures are out of control. Power will pass to the Executive. Impoundment *de facto* will be replaced by impoundment *de jure*, for everyone will know that without fiscal responsibility there can be no financial power. Since future increases are built into present budgets—increases in social security benefits may well take up all the slack for over a decade unless there are to be huge deficits—Congressional restraint in the face of desire for social programs may collapse.

Parties would not be anyone's current "man-on-a-horse" to control Presidents. The Republican National Committee avoided implication in Watergate because President Nixon was able to brush it aside in order to run his own campaign. Nevertheless, the party must accept a large share of the blame. If party leaders may be excused for not stopping Watergate before it got started, they must surely be faulted for not halting the cover-up once it became visible. The remarkable thing was the President's perseverance in the pattern of secret . . . forced revelation . . . new secret . . . new forced revelation . . . ad nauseam. Had there been party leaders (Congressmen, national committee members, governors, state chairmen, city "bosses") with power in their own jurisdictions, they could have descended on the President en masse and said they would denounce him if he did not do the right thing. They, at least, had to be told the truth so they could judge the truth to tell. But this scenario did not occur. Why not?

The nation lacks party leaders because its parties are weak. The same weakness that allows a Barry Goldwater or a George McGovern, candidates without substantial popular support, to take the nomination for President also makes party leaders ineffective in bringing pressure to bear on their President. At a time when the nation most needs the restraint on office holders exercised by organizations with an interest in the long-term repute of their party, it can no longer call on them. The price of making parties too weak to do harm turns out to be rendering party leaders too useless to do good.

What is in it for anyone to be a hardworking member of a political party? Except for the few who go on to public office, all the party activist can expect is abuse. What used to be candidate selection is now determined in primaries; what used to be patronage is now called civil service; what used to be status has turned into sneer. The politician has replaced the rich man as he who will as soon enter heaven as a camel passeth through the eye of a needle.

Reform, which one might think would strengthen parties, is in fact designed to weaken them. Once a candidate gets financial support from the government, he has less use not only for "fat cat" financiers but for "lean kitten" party politicians. The latest version of "The Incumbent Protection Act" will not make legislators more amenable to party considerations. New rules for party conferences and conventions stress expression at the expense of election. It is as if the Democratic Party did not know itself but had to discover who it was at the last minute by descending into the streets. Instead of being a place where the party meets to choose candidates who can win elections, conventions become a site for expressing the delegates' awareness of who they are. The politics of existentialism replaces the existence of politics.

The demise of parties is paralleled by the rise of citizen lobbies like Common Cause whose main thrust is to weaken other intermediary organizations—parties, labor unions, trade associations—that stand between citizen and government. Largely middle-class and upper-middle-class in composition, these "public interest" lobbies seek to reduce the influence of "private interests" by limiting the money they can contribute or the activities they can carry on to mold politicians or shape legislation. The sun that these laws are supposed to bring to politics, however, does not shine equally on all classes. Without strong unions and parties, workers will find

that "open" hearings will, in effect, be "closed" to them. Corporations will continue to use lobbyists but they can pack no meetings.
Only citizen lobbies combine the cash and the flexible hours of
middle-class professionals needed to produce "mass" mobilization
aimed at specific targets. Eventually, as intermediary groups decline, the idea will grow that lively citizen lobbies should replace
moribund political parties. But as the lobbies weaken the capacity
of social interests to bargain, they will discover that they cannot
perform the integrative function of parties.

Politics in the liberal society

If party and legislature cannot constrain Presidents, can the press
do the job? It can do its jobs—to expose and to ventilate—but it
cannot do the President's job: to provide leadership in critical areas
of public policy. So long as people are concerned with abuse of
power, the press will be powerful. Its essentially negative role is
then viewed in a positive light. But when the pervasive problem
in government is lack of power, the critical virtues of the media
of information will become their carping vices.

Today we have become accustomed to certain constants: Presidents are rarely satisfied with their portrayal in the media of information, and reporters and commentators constantly complain
about attempts to mislead or control them. They never seem to
find a President who loves public criticism; he never seems to find
a press and television that love him. It is in the President's interest
to put the best gloss he can on things, and it is the reporter's vo-
cation to find out what is "really" happening. Presidents are judged
on the news and they can't help wishing it were good. But the
news cannot be good when conditions are bad. No President in
the near future will get good grades and every President will want
to flunk the press.

The media, after all, are part of society. They reflect the dis-
satisfactions with government that grew in the 1960's. Scoops used
to go to reporters who played along; today Pulitzer prizes go to
reporters who get things they are not supposed to have. The leak
of the Pentagon Papers, which did so much to galvanize the Nixon
Administration in precisely the wrong way, would not have been
possible without a public opinion to which challenging government had become infinitely more respectable than it was a decade
before. Any reporter in Britain, whatever his party and policy preferences, is more a part of the permanent government (the elites

that rule this time or expect to next time) than is any reporter in
America, even if his favorite occupies the White House. Newsmen
are not merely observers; they are also, in David S. Broder's apt
title, "Political Reporters in Presidential Politics."

The perennial quarrel between the press and the President inadvertently points up an aspect of their relationship that is more
important than whether one institution gets along with the other:
Like everyone else in America, the press is fascinated by the Presidency. If its exaltation of the incumbent has declined, its fixation
on him has not. Whom the press talks about is as important as
what it says (if not more so). The media reinforce the identification of the Presidency with the political system.

People generally ask why their Presidents are doing so little rather than so much. Since America is a liberal society, it is not surprising that its people developed a liberal theory of the Presidency. All power to the Presidency? We never quite went that far.
What is good for the Presidency is good for the country—we came
near to that. The people wanted a New Deal, and a President gave
it to them. They wanted egalitarian social legislation and, on the
whole, Presidents were disposed to give it to them. The people
wanted more, and Presidents could promise more. They wanted
novelty, and Presidents provided it—new frontiers, great societies,
fair deals, crusades. . . .

By comparison, Congress appeared confused, courts seemed
mired in precedent, and bureaucracies were tangled in red tape.
Checks turned into obstructions and balances became dead weights.
The separation of powers looked like an 18th-century anachronism
in a 20th-century world. In short, we were concerned with results,
and unconcerned with institutions. The idea that the Presidency
was implicated in a system of checks and balances, that the safety
of the nation lay not in individual virtue but in institutional arrangements, was, though given lip service, in fact given short
shrift. Once again, soon enough, the people will want a strong,
action-oriented President to override obstructive courts and dila-
tory and divided Congresses.

Bureaucratic frustrations

The real rival to the Presidency will be the bureaucracy, if not
by intent, then through inadvertence. It is not so much that bureaucrats might resist a Presidential lead, though they might, but
that such leads will be increasingly hard to give. The sheer size

of government means that the Executive Office inevitably knows less about what is going on and what to do about it than it has in the past. Nor can any one person, including the President, know more than a small proportion of what his staff knows, which in turn is only part of the picture. The more frequently that government tries to intervene more deeply in society, the less anyone is able to control events and further removed the Presidency is bound to be from what is actually going on. When government is over-burdened, bureaucracy becomes unbearable; the march of complexity is mainly responsible for the mounting frustration of Presidents with "their" bureaucrats.

When Richard Fenno first published his classic study *The President's Cabinet* in 1959, it appeared that this cadre was at its lowest ebb. Its lack of a collective interest in decisions made by others and its dependence on the President rather than on Cabinet colleagues explained why it had never really been strong in the history of American politics. Under Presidents Kennedy, Johnson, and Nixon, however, the Cabinet appeared not merely weak but virtually non-existent. Its former low ebb seems, in retrospect, almost to have been its high tide.

What happened? In the old days, Presidents mistrusted the Cabinet because they were not entirely free in choosing its members. They had, in effect, to "give away" appointments to important party factions outside the government and to powerful Congressmen inside. The decline of party in American political life, however, has meant that recent Presidents have not been beholden to it. The increasing diffusion of power within Congress has also meant that there were correspondingly fewer powerful leaders whom Presidents have had to cater to in making appointments. Yet their increased freedom in selecting Cabinet members apparently has not led Presidents to confer greater trust on their appointees. Why not?

Secretaries of the great departments must serve more than one master. They are necessarily beholden to Congress for appropriations and for substantive legislation. They are expected to speak for the major interests entrusted to their care, as well as for the President. They need cooperation from the bureaucracy that surrounds them, and they may have to make accommodations to get that support. A Secretary of Agriculture who is vastly unpopular with farmers, a Secretary of Interior who is hated by conservationists, and all Secretaries whose employees undermine their efforts, cannot be of much use to the Chief Executive. Nevertheless, Presidents (and especially their personal staffs) appear to behave as if there were something wrong when Cabinet members do what comes naturally to people in their positions.

To the White House staff the separation of powers is anathema. They have wonderful ideas, apparently, only to see them sabotaged in the bureaucratic labyrinth. How dare those bureaucrats get in the way! The notion that the departments might owe something to Congress or that there is more to policy than what the President and his men want flickers only occasionally across their minds. It is as if the Presidency were THE government. The President's men tend to see themselves not as part of a larger system, but as the system itself.

Alternatively, going to the other extreme, consider the image of a beleaguered outpost in the White House with all the insignia of office but without the ability to command troops in the provinces. As Arthur Schlesinger, Jr., put it when reflecting on his experiences as an adviser to President Kennedy in *A Thousand Days,* "Wherever we have gone wrong . . . has been because we have not had sufficient confidence in the New Frontier approach to impose it on the government. Every important mistake has been the consequence of excessive deference to the permanent government. . . . " Small wonder, then, that Presidents in later times have interpreted discontent with their policies or their behavior not as evidence of the intractability of the policy problems themselves but as another effort by the "big bad bureaucracy" (readily expanded to include the "establishment press") to frustrate the people.

Courts and the curbing of power

The protection afforded by the courts in the Watergate case really shows how limited their role must be. If we are to depend on future criminal actions to control Presidents, if they can do anything they please that isn't plainly illegal, then they can do "most anything." It is no derogation of the courts to say that they are (with a few notable exceptions) reactive. They may tell others what to do after (usually a long time after) the fact, but they cannot compete with executives and legislatures for control of future choices. Indeed, efforts to alter future behavior by writing into law (or interpreting) prohibitions against specific past events are bound to fail.

The legal profession has the right aphorism: Hard cases make bad law. Congress has, after the fact, passed the War Powers Act to restrict Presidents in the future. In my opinion, the gesture was futile and possibly dangerous. Basically, the act provides that within 60 days of the commencement of hostilities the President must receive positive Congressional approval, or the military action grinds to a halt. Would this law have stopped Vietnam? That is doubtful because of the spirit prevailing at the time of the Gulf of Tonkin Resolution. Will it enable a President to respond when necessary and restrain him when essential? Such judgments, I fear, are more readily made in retrospect than in prospect. My concern is that the War Powers Act can be converted too easily into a permanent Gulf of Tonkin Resolution bestowing a legislative benediction in advance on Presidential actions, by which time Congress has little choice left. Now Presidents can point to a statute saying that they can do what they want (the Mayaguez affair should dispose of the requirement for consultation) for up to 60 days. The lesser defect lies in trying to anticipate the specific configuration of events by general statutory principles. The greater defect lies in trying to frame general principles on the basis of the most recent horrible event. And the greatest defect resides in treating a systemic problem, whose resolution depends on the interaction of numerous parts in a dynamic environment, as if it were a defect in a single component operating in a static situation.

History shows, to be sure, that constitutions are often written against the last usurper—the United States Constitution against the Articles of Confederation, the Fifth French Republic against the Fourth and the Third, Bonn against Weimar, the United Nations contra the League of Nations. So, too, do generals frequently end up preparing for the last war. But this is not quite so bad as aiming living constitutional provisions at dead targets. The nation has not been well served by the anti-third-term (read anti-Franklin Roosevelt) amendment, which guarantees that if the people ever find a President they deem worth keeping in office, they will have prohibited themselves from doing so. In the guise of denying power to Presidents, it has, in fact, been denied to the people. Were it not for the two-term limitation, moreover, we would be spared the current nonsense of proposing a single six-year term. Why some people think popular control over Presidents would be increased by taking them out of the electoral arena after they assume office must remain a mystery.

The difficulties judges face in trying to capture a probabilistic process of institutional interaction in a deterministic principle is nowhere better illustrated than in the recent pronouncement of the Supreme Court on executive privilege. In the old days an assertion of executive privilege had no legal status. No one knew what it meant, which was just the way things should be. Presidents needed confidentiality; Congresses needed information; these self-evident truths were left to contend with each other according to the circumstances of the time and the pulling and hauling of the participants. Eventually an accommodation was reached that lasted until the question was raised again under altered conditions. Richard Nixon changed all that by bending the principle so far that it broke in his hands. Although the Supreme Court could not quite put executive privilege back together again, it went one step beyond by giving legal sanction to a doctrine that never had one. In order to achieve a unanimous opinion that President Nixon had to give up the tapes, the Supreme Court apparently felt it necessary to distinguish this case from all others by declaring that executive privilege might exist in other cases but manifestly not in this one. For the first time, future Presidents will be able to cite an opinion to the effect that there is a legal basis for executive privilege.

Suppose all the other institutions—Congress, courts, bureaucracies, parties—ganged up on the Presidency: Couldn't they curb Presidential power? No doubt they could. But the likelihood of their all getting together is small, unless they face a common threat. Now what President would be so foolish as to attack each and every other institution all at the same time? Nixon, of course —the House on expenditures, the Senate on foreign policy, the media on misinformation, the courts on credibility, and on and on. Are we then to think of Watergate as the modal condition of the Presidency? No, it is more like a limiting condition: Many things have to go wrong at once before the Presidency enters into a knockabout against all comers. If future Presidents have to do all that Nixon did before their wings are clipped, they will win every race against their institutional competitors.

There are, to be sure, contrary trends that might lead to the emergence of demagogic Presidents who would hold sway over the multitudes. The decline in party identification leaves latitude for personality differences among the candidates to show up in the form of landslide majorities. And the growing influence of issue enthusiasts in national nominating conventions means that the major parties are more likely to nominate candidates preferred by their extremes but rejected by the population at large. This is what happened to the Goldwater Republicans in 1964 and the McGovern Democrats in 1972. But the popularity of the winners, Johnson and Nixon, did not last long.

It is also possible that the very absence of consensus on foreign policy and the very presence of contradictions at the heart of domestic policy would lead to a call for a leader to overcome (or more accurately, suppress) these disturbing conditions. Under these circumstances, threats to liberty are likely to arise less from a desire for personal Presidential aggrandizement, and more from mass insistence that power be exercised to eliminate ambiguity. The Presidential problem will not be power but performance.

The flight of the Presidency

Perhaps the most interesting events of our time are those that have not occurred—the failure of demagogues, parties, or mass movements to take advantage of the national disarray. Maybe the country is in better shape than we think, or they (whoever "they" are) are waiting for things to get still worse. In any event, it now appears that the United States will have to get out of its predicaments with the same ordinary, everyday, homespun political institutions with which it got into them. How will its future leaders (preeminently its Presidents) appraise the political context in which they find themselves? For leadership is not a unilateral imposition but a mutual relationship, in which it is as important to know whether and wither people are willing to be led as why and where their leaders propose to take them.

When you bite the hand that feeds you, it moves out of range. That is the significance of the near-geometric increase in the size of the Executive Office from Truman's to Nixon's time. Originally a response to the growth of government, it became a means of insulating Presidents from the shocks of a society with which they could no longer cope.

As government tried to intervene both more extensively and more intensively in the lives of citizens, the executive office, in order to monitor these events, became a parallel bureaucracy. The larger it grew and the more programs it tried to cover, however, the more the Executive Office inadvertently but inevitably became further removed from the lives of the people who were being affected by these new operations. Then the inevitable became desirable and the inadvertent became functional; for as presidents discovered that domestic programs paid no dividends—see Nixon's comment in the Watergate tapes about the futility of building more outhouses in Peoria—insulation from the people didn't seem such a bad idea.

After Watergate, it became necessary to reduce the size of the

Executive Office and to increase the number of people reporting directly to the President, so as to show he was as different as possible from his predecessor. Disassociation from Watergate, however, is not the same as organization for action. Faced with a lack of consensus in foreign policy and an absence of support for domestic policy, future Presidents may well retreat in the face of overwhelming odds.

In the first months of Richard Nixon's second term, when his Administration was at full strength and he was attempting to chart a new course, he tried both to reorganize radically the federal bureaucracy and to alter drastically his own relations to it. In part, the idea was old—rationalize the bureaucracy by creating a smaller number of bigger departments, thus cutting down on the large number of people previously entitled to report directly to the President. Part of it was new—creating a small group of supersecretaries who would have jurisdiction over several departments, thus forming, in effect, an inner Cabinet. Part of it was peculiarly Nixonian—drastically reducing demands on his time and attention. Part of it, I believe, is likely to be permanent—adapting to the increasing scope and complexity of government by focusing Presidential energy on a few broadly defined areas of policy. Presidents have to deal with war and peace and with domestic prosperity or lack of it. There will continue to be a person the President relies on principally for advice in foreign and defense policy, and another on whom he relies for economic management. The demands on Presidents for response to other domestic needs vary with the times—so that there will undoubtedly be one or two other superadvisers, called by whatever names or holding whatever titles, to deal with race relations or energy or the environment or whatever else seems most urgent. The rest will be a residual category—domestic policy supervised by a domestic council—reserved for a Vice President or a Cabinet Coordinator.

The response to ever-increasing complexity will continue to be ever-greater simplicity. This is the rationale behind wholesaling instead of retailing domestic policies; behind revenue sharing instead of endless numbers of categorical grants; behind proposals for family assistance and negative income taxes instead of a multiplicity of welfare policies; behind a transfer to state and local governments of as much responsibility (though not necessarily as much money) as they can absorb. "Here is a lot of trouble and a little money," these Presidential policies seem to say, "so remember the trauma is all yours and none of mine."

The "offensive retreat"

Future Presidents will be preoccupied with operating strategic levers, not with making tactical moves. They will see their power stakes, to use Neustadt's term, in giving away their powers; like everyone else they will have to choose between what they have to keep and what they must give up. Not so much running the country (that was Nixon's error) but seeing that it is running will be their forte. The Cabinet, or at least the inner "Super-Secretary" Cabinet, will undergo a visible revival because Presidents will trade a little power for a lot of protection. The more prominent a President's Cabinet, the less of a target he becomes. When Presidents wanted to keep the credit, they kept their Cabinets quiet; but they will welcome Cabinet notoriety now that they want to spread the blame.

The "offensive retreat" of the Presidency will not be the work of a single President or a particular moment in time. Nor will this movement be unidirectional. Like most things, it will be a product of trial and error in which backsliding will be as prominent as forward movement. But as Presidents discover there is sickness in health and ignorance in education, they will worry more about their own welfare.

Presidents, and the governments for which they stand, are either doing too much or too little. They need either to do a great deal more for the people or a great deal less to them. They must be closer to what is happening or much further away. At present, they are close enough to get the blame but too far away to control the result; for government to be half involved is to be wholly abused. Which way will it move?

One way is nationalization. The federal government would take over all areas of serious interest; there would be a National Health Service, a National Welfare System, and the like. Industry would not be regulated at a remove—but run for real. The blame would go to the top but so would the power. Presidents would literally be running the country. The danger of overload at the center would be mitigated by mastery of the periphery. Uncle Sam would be everybody's tough Uncle and it would not be wise to push him around. Before it comes to this, however, Presidents will try to move in other directions.

Presidents will seek the fewest levers but those with the most consequential effects. They will be money men, manipulating the supply to citizens through income floors and the supply to business through banks. Taxes will vary with expenditures; if government spends more, taxes will go up, and if it spends less, they will go down, with the upper limit set by constitutional limitation. Income will determine outgo. Regulatory activities and agencies will be severed from all Presidential connection; why should Presidents get into trouble for fixing the price of milk or determining routes for airlines or setting railroad rates? States and localities will undertake whatever supplementary programs they are willing and able to support.

Presidents will handle systemic crises, not ordinary events. They will be responsible for war and peace abroad, and life and death at home, but not much in between. The people will not call on their President when they are in trouble. The President will call on his people when he is in trouble, for Presidents will represent the general interest in maintaining essential services and institutions, not the private and personal interests of every individual in his and her own fate.

Presidential prospects

The framers of our Constitution intended that its overarching structure should restrain each institution through the mutual interaction of its parts. Citizens should now see that their safety lies in that restraint. So long as the Presidency is seen as part of the system, subject to its checks and balances, citizens retain hope and show calm. The institutional lesson to be learned, therefore, is not that the Presidency should be diminished but that other institutions should grow in stature. The first order of priority should go to rebuilding our political parties, because they are most in need of help and could do most to bring Presidents in line with strong sentiments in the country. Had there been Republican "elders" of sufficient size and weight, the President representing their party could not have so readily strayed in his perception of the popular will. Whatever its other defects, a party provides essential connective tissue between people and government. So do the media. So does Congress; strengthening its appropriations process through internal reform would bring it more power than any external threat could take away. The people need the vigor of all their institutions.

Strengthening the political system as a whole will not necessarily weaken the Presidency. On the other hand, making the Presidency into the government would not only threaten liberty but leave it without support in times of adversity. No one wants to be president of a bankrupt company; Chairman of the board of a going concern is more like it. The national objective should be to increase total systemic capacity in relation to emerging problems. One way to do this is to lower people's expectations of what government and its leaders can do for them. Maybe that is what Watergate has tried to tell us. Another way is to depend on leadership. Maybe that is what President Ford's pardon of his predecessor should teach us not to do.

For 30 halcyon days the people of the United States had a President they could trust. This hope and trust was a precious national resource. Its dissipation was a national calamity. For it did not belong to Ford alone but was bestowed upon him by virtue of the circumstance under which he entered office. His task was to husband and preserve it for the dark days ahead. This was not only his but also the nation's most precious stock of political capital and he squandered it with a sudden wastefulness. If it were a question of the former President suffering more and the people less, Ford clearly mistook his priorities.

The voice of Ezekiel—"Son of man, trust not in man"—has a contemporary echo. If only Nixon went, the hope was, a leader would arise among us. Apparently not. The resignation and pardon warn against passive dependence on leaders. The wisdom of a democracy must lie in its "separated institutions sharing power." The virtues of a democracy must ultimately be tested, not in the leadership of its fallible men, but in the enduring power of its great institutions.

But institutions are not everything. They function in a climate of opinion that both limits and shapes what they can do. Those who operate them respond to the rewards and punishments of the environment in which they are situated. A people who punish truthfulness will get lies; a people who reward symbolic actions will get rhetoric instead of realism. So long as the people appear to make contradictory demands for domestic policies, they will be supplied by contrary politicians. Until the country is prepared to support a foreign policy that is responsive to new events, it will continue to nourish old misunderstandings.

Long before the current disenchantment, Henry Fairlie, in his book *The Kennedy Promise*, had questioned whether Americans did not have too exalted an opinion of what politics (and hence Presidents) could do. To some, Kennedy's Camelot recedes in the shimmering distance as nobility thwarted, its light tragically extinguished before its time. Fairlie discerned in the Kennedys an excessive conception of what government could do, a sense of pol-

itics over society that would come to no good. As he puts it, "The
people are encouraged to expect too much of their political institutions and of their political leaders. They cease to inquire what
politics may accomplish for them, and what they must do for themselves. Instead, they expect politics to take the place religion once
held in their lives. ..." When leaders were gods, it is worth re-
calling, they punished their people.

Public officials need to know they cannot lead by following a
people that does not know where it wants to go, or how to get
there. They must persuade themselves—and, in so doing, the people they wish to lead—that no one can have it all or at the same
time. As officials seek to improve their public performance, they
must simultaneously strive to shape popular expectations, for unless the two meet there can be no hope for (and perhaps no distinction between) the leaders and the led. If they fail, leaders will
move not with but from their people. If they succeed, then the
short-term predictions of this essay—the flight of the Presidency
from its people—need not become the longer-run prognostication
for the larger political system.

4

The
rise of the
bureaucratic
state

JAMES Q. WILSON

URING its first 150 years, the American republic was not thought to have a "bureaucracy," and thus it would have been meaningless to refer to the "problems" of a "bureaucratic state." There were, of course, appointed civilian officials: Though only about 3,000 at the end of the Federalist period, there were about 95,000 by the time Grover Cleveland assumed office in 1881, and nearly half a million by 1925. Some aspects of these numerous officials were regarded as problems—notably, the standards by which they were appointed and the political loyalties to which they were held—but these were thought to be matters of proper character and good management. The great political and constitutional struggles were not over the power of the administrative apparatus, but over the power of the President, of Congress, and of the states.

The Founding Fathers had little to say about the nature or function of the executive branch of the new government. The Constitution is virtually silent on the subject and the debates in the Constitutional Convention are almost devoid of reference to an administrative apparatus. This reflected no lack of concern about the matter, however. Indeed, it was in part because of the Founders' depressing experience with chaotic and inefficient management un-

der the Continental Congress and the Articles of Confederation
that they had assembled in Philadelphia. Management by committees composed of part-time amateurs had cost the colonies dearly
in the War of Independence and few, if any, of the Founders wished
to return to that system. The argument was only over how the heads
of the necessary departments of government were to be selected,
and whether these heads should be wholly subordinate to the President or whether instead they should form some sort of council that
would advise the President and perhaps share in his authority. In
the end, the Founders left it up to Congress to decide the matter.

There was no dispute in Congress that there should be executive
departments, headed by single appointed officials, and, of course,
the Constitution specified that these would be appointed by the
President with the advice and consent of the Senate. The only issue
was how such officials might be removed. After prolonged debate
and by the narrowest of majorities, Congress agreed that the President should have the sole right of removal, thus confirming that
the infant administrative system would be wholly subordinate—in
law at least—to the President. Had not Vice President John Adams,
presiding over a Senate equally divided on the issue, cast the deciding vote in favor of Presidential removal, the administrative departments might conceivably have become legal dependencies of
the legislature, with incalculable consequences for the development
of the embryonic government.

The "bureaucracy problem"

The original departments were small and had limited duties. The
State Department, the first to be created, had but nine employees
in addition to the Secretary. The War Department did not reach 80
civilian employees until 1801; it commanded only a few thousand
soldiers. Only the Treasury Department had substantial powers—it
collected taxes, managed the public debt, ran the national bank,
conducted land surveys, and purchased military supplies. Because
of this, Congress gave the closest scrutiny to its structure and its
activities.

The number of administrative agencies and employees grew slowly but steadily during the 19th and early 20th centuries and then
increased explosively on the occasion of World War I, the Depression, and World War II. It is difficult to say at what point in this
process the administrative system became a distinct locus of power or an independent source of political initiatives and problems.

What is clear is that the emphasis on the sheer *size* of the administrative establishment—conventional in many treatments of the subject—is misleading.

The government can spend vast sums of money—wisely or unwisely—without creating that set of conditions we ordinarily associate with the bureaucratic state. For example, there could be massive transfer payments made under government auspices from person to person or from state to state, all managed by a comparatively small staff of officials and a few large computers. In 1971, the federal government paid out $54 billion under various social insurance programs, yet the Social Security Administration employs only 73,000 persons, many of whom perform purely routine tasks.

And though it may be harder to believe, the government could in principle employ an army of civilian personnel without giving rise to those organizational patterns that we call bureaucratic. Suppose, for instance, that we as a nation should decide to have in the public schools at least one teacher for every two students. This would require a vast increase in the number of teachers and school rooms, but almost all of the persons added would be performing more or less identical tasks, and they could be organized into very small units (e.g., neighborhood schools). Though there would be significant overhead costs, most citizens would not be aware of any increase in the "bureaucratic" aspects of education—indeed, owing to the much greater time each teacher would have to devote to each pupil and his or her parents, the citizenry might well conclude that there actually had been a substantial reduction in the amount of "bureaucracy."

To the reader predisposed to believe that we have a "bureaucracy problem," these hypothetical cases may seem farfetched. Max Weber, after all, warned us that in capitalist and socialist societies alike, bureaucracy was likely to acquire an "overtowering" power position. Conservatives have always feared bureaucracy, save perhaps the police. Humane socialists have frequently been embarrassed by their inability to reconcile a desire for public control of the economy with the suspicion that a public bureaucracy may be as immune to democratic control as a private one. Liberals have equivocated, either dismissing any concern for bureaucracy as reactionary quibbling about social progress, or embracing that concern when obviously nonreactionary persons (welfare recipients, for example) express a view toward the Department of Health, Education, and Welfare indistinguishable from the view businessmen take of the Internal Revenue Service.

Political authority

There are at least three ways in which political power may be gathered undesirably into bureaucratic hands: by the growth of an administrative apparatus so large as to be immune from popular control, by placing power over a governmental bureaucracy of any size in private rather than public hands, or by vesting discretionary authority in the hands of a public agency so that the exercise of that power is not responsive to the public good. These are not the only problems that arise because of bureaucratic organization. From the point of view of their members, bureaucracies are sometimes uncaring, ponderous, or unfair; from the point of view of their political superiors, they are sometimes unimaginative or inefficient; from the point of view of their clients, they are sometimes slow or unjust. No single account can possibly treat of all that is problematic in bureaucracy; even the part I discuss here—the extent to which political authority has been transferred undesirably to an unaccountable administrative realm—is itself too large for a single essay. But it is, if not the most important problem, then surely the one that would most have troubled our Revolutionary leaders, especially those that went on to produce the Constitution. It was, after all, the question of power that chiefly concerned them, both in redefining our relationship with England and in finding a new basis for political authority in the Colonies.

To some, following in the tradition of Weber, bureaucracy is the inevitable consequence and perhaps necessary concomitant of modernity. A money economy, the division of labor, and the evolution of legal-rational norms to justify organizational authority require the efficient adaptation of means to ends and a high degree of predictability in the behavior of rulers. To this, Georg Simmel added the view that organizations tend to acquire the characteristics of those institutions with which they are in conflict, so that as government becomes more bureaucratic, private organizations—political parties, trade unions, voluntary associations—will have an additional reason to become bureaucratic as well.

By viewing bureaucracy as an inevitable (or, as some would put it, "functional") aspect of society, we find ourselves attracted to theories that explain the growth of bureaucracy in terms of some inner dynamic to which all agencies respond and which makes all barely governable and scarcely tolerable. Bureaucracies grow, we are told, because of Parkinson's Law: Work and personnel expand to consume the available resources. Bureaucracies behave, we believe, in accord with various other maxims, such as the Peter Principle:

In hierarchical organizations, personnel are promoted up to that point at which their incompetence becomes manifest—hence, all important positions are held by incompetents. More elegant, if not essentially different, theories have been propounded by scholars. The tendency of all bureaus to expand is explained by William A. Niskanen by the assumption, derived from the theory of the firm, that "bureaucrats maximize the total budget of their bureau during their tenure"—hence, "all bureaus are too large." What keeps them from being not merely too large but all-consuming is the fact that a bureau must deliver to some degree on its promised output, and if it consistently underdelivers, its budget will be cut by unhappy legislators. But since measuring the output of a bureau is often difficult—indeed, even *conceptualizing* the output of the State Department is mind-boggling—the bureau has a great deal of freedom within which to seek the largest possible budget.

Such theories, both the popular and the scholarly, assign little importance to the nature of the tasks an agency performs, the constitutional framework in which it is embedded, or the preferences and attitudes of citizens and legislators. Our approach will be quite different: Different agencies will be examined in historical perspective to discover the kinds of problems, if any, to which their operation gave rise, and how those problems were affected—perhaps determined—by the tasks which they were assigned, the political system in which they operated, and the preferences they were required to consult. What follows will be far from a systematic treatment of such matters, and even farther from a rigorous testing of any theory of bureaucratization: Our knowledge of agency history and behavior is too sketchy to permit that.

Bureaucracy and size

During the first half of the 19th century, the growth in the size of the federal bureaucracy can be explained, not by the assumption of new tasks by the government or by the imperialistic designs of the managers of existing tasks, but by the addition to existing bureaus of personnel performing essentially routine, repetitive tasks for which the public demand was great and unavoidable. The principal problem facing a bureaucracy thus enlarged was how best to coordinate its activities toward given and noncontroversial ends.

The increase in the size of the executive branch of the federal government at this time was almost entirely the result of the increase in the size of the Post Office. From 1816 to 1861, federal civilian employment in the executive branch increased nearly eight-fold (from 4,837 to 36,672), but 86 per cent of this growth was the result of additions to the postal service. The Post Office Department was expanding as population and commerce expanded. By 1869 there were 27,000 post offices scattered around the nation; by 1901, nearly 77,000. In New York alone, by 1894 there were nearly 3,000 postal employees, the same number required to run the entire federal government at the beginning of that century.

The organizational shape of the Post Office was more or less fixed in the administration of Andrew Jackson. The Postmaster General, almost always appointed because of his partisan position, was aided by three (later four) assistant postmaster generals dealing with appointments, mail-carrying contracts, operations, and finance. There is no reason in theory why such an organization could not deliver the mails efficiently and honestly: The task is routine, its performance is measurable, and its value is monitored by millions of customers. Yet the Post Office, from the earliest years of the 19th century, was an organization marred by inefficiency and corruption. The reason is often thought to be found in the making of political appointments to the Post Office. "Political hacks," so the theory goes, would inevitably combine dishonesty and incompetence to the disservice of the nation; thus, by cleansing the department of such persons these difficulties could be avoided. Indeed, some have argued that it was the advent of the "spoils system" under Jackson that contributed to the later inefficiencies of the public bureaucracy.

The opposite is more nearly the case. The Jacksonians did not seek to make the administrative apparatus a mere tool of the Democratic party advantage, but to purify that apparatus not only of what they took to be Federalist subversion but also of personal decadence. The government was becoming not just large, but lax. Integrity and diligence were absent, not merely from government, but from social institutions generally. The Jacksonians were in many cases concerned about the decline in what the Founders had called "republican virtue," but what their successors were more likely to call simplicity and decency. As Matthew Crenson has recently observed in his book *The Federal Machine,* Jacksonian administrators wanted to "guarantee the good behavior of civil servants" as well as to cope with bigness, and to do this they sought both to place their own followers in office and—what is more important—to create a system of depersonalized, specialized bureaucratic rule. Far from being the enemies of bureaucracy, the Jacksonians were among its principal architects.

Impersonal administrative systems, like the spoils system, were "devices for strengthening the government's authority over its own civil servants"; these bureaucratic methods were, in turn, intended to "compensate for a decline in the disciplinary power of social institutions" such as the community, the professions, and business. If public servants, like men generally in a rapidly growing and diversifying society, could no longer be relied upon "to have a delicate regard for their reputations," accurate bookkeeping, close inspections, and regularized procedures would accomplish what character could not.

Amos Kendall, Postmaster General under President Jackson, set about to achieve this goal with a remarkable series of administrative innovations. To prevent corruption, Kendall embarked on two contradictory courses of action: He sought to bring every detail of the department's affairs under his personal scrutiny and he began to reduce and divide the authority on which that scrutiny depended. Virtually every important document and many unimportant ones had to be signed by Kendall himself. At the same time, he gave to the Treasury Department the power to audit his accounts and obtained from Congress a law requiring that the revenues of the department be paid into the Treasury rather than retained by the Post Office. The duties of his subordinates were carefully defined and arranged so that the authority of one assistant would tend to check that of another. What was installed was not simply a specialized management system, but a concept of the administrative separation of powers.

Few subsequent postmasters were of Kendall's ability. The result was predictable. Endless details flowed to Washington for decision but no one in Washington other than the Postmaster General had the authority to decide. Meanwhile, the size of the postal establishment grew by leaps and bounds. Quickly the department began to operate on the basis of habit and local custom: Since everybody reported to Washington, in effect no one did. As Leonard D. White was later to remark, "the system could work only because it was a vast, repetitive, fixed, and generally routine operation." John Wanamaker, an able businessman who became Postmaster General under President Cleveland, proposed decentralizing the department under 26 regional supervisors. But Wanamaker's own assistants in Washington were unenthusiastic about such a diminution in their authority and, in any event, Congress steadfastly refused to endorse decentralization.

Civil service reform was not strongly resisted in the Post Office;

from 1883 on, the number of its employees covered by the merit system expanded. Big-city postmasters were often delighted to be relieved of the burden of dealing with hundreds of place-seekers. Employees welcomed the job protection that civil service provided. In time, the merit system came to govern Post Office personnel almost completely, yet the problems of the department became, if anything, worse. By the mid-20th century, slow and inadequate service, an inability technologically to cope with the mounting flood of mail, and the inequities of its pricing system became all too evident. The problem with the Post Office, however, was not omnipotence but impotence. It was a government monopoly. Being a monopoly, it had little incentive to find the most efficient means to manage its services; being a government monopoly, it was not free to adopt such means even when found—communities, Congressmen, and special-interest groups saw to that.

The military establishment

Not all large bureaucracies grow in response to demands for service. The Department of Defense, since 1941 the largest employer of federal civilian officials, has become, as the governmental keystone of the "military-industrial complex," the very archetype of an administrative entity that is thought to be so vast and so well-entrenched that it can virtually ignore the political branches of government, growing and even acting on the basis of its own inner imperatives. In fact, until recently the military services were a major economic and political force only during wartime. In the late 18th and early 19th centuries, America was a neutral nation with only a tiny standing army. During the Civil War, over two million men served on the Union side alone and the War Department expanded enormously, but demobilization after the war was virtually complete, except for a small Indian-fighting force. Its peacetime authorized strength was only 25,000 enlisted men and 2,161 officers, and its actual strength for the rest of the century was often less. Congress authorized the purchase and installation of over 2,000 coastal defense guns, but barely six per cent of these were put in place.

When war with Spain broke out, the army was almost totally unprepared. Over 300,000 men eventually served in that brief conflict, and though almost all were again demobilized, the War Department under Elihu Root was reorganized and put on a more professionalized basis with a greater capacity for unified central control. Since the United States had become an imperial power with important possessions in the Caribbean and the Far East, the need for a larger military establishment was clear; even so, the average size of the army until World War I was only about 250,000.

The First World War again witnessed a vast mobilization—nearly five million men in all—and again an almost complete demobilization after the war. The Second World War involved over 16 million military personnel. The demobilization that followed was less complete than after previous engagements, owing to the development of the Cold War, but it was substantial nonetheless—the Army fell in size from over eight million men to only half a million. Military spending declined from $91 billion in the first quarter of 1945 to only slightly more than $10 billion in the second quarter of 1947. For the next three years it remained relatively flat. It began to rise rapidly in 1950, partly to finance our involvement in the Korean conflict and partly to begin the construction of a military force that could counterbalance the Soviet Union, especially in Europe.

In sum, from the Revolutionary War to 1950, a period of over 170 years, the size and deployment of the military establishment in this country was governed entirely by decisions made by political leaders on political grounds. The military did not expand autonomously, a large standing army did not find wars to fight, and its officers did not play a significant potential role except in wartime and occasionally as Presidential candidates. No bureaucracy proved easier to control, at least insofar as its size and purposes were concerned.

A "military-industrial complex"?

The argument for the existence of an autonomous, bureaucratically-led military-industrial complex is supported primarily by events since 1950. Not only has the United States assumed during this period world-wide commitments that necessitate a larger military establishment, but the advent of new, high-technology weapons has created a vast industrial machine with an interest in sustaining a high level of military expenditures, especially on weapons research, development, and acquisition. This machine, so the argument goes, is allied with the Pentagon in ways that dominate the political officials nominally in charge of the armed forces. There is some truth in all this. We have become a world military force, though that decision was made by elected officials in 1949-1950 and not dictated by a (then nonexistent) military-industrial complex. High-cost, high-technology weapons have become important and a number of industrial concerns will prosper or perish depending on how contracts for those weapons are let. The development and purchase of weapons is sometimes made in a wasteful, even irrational, manner. And the allocation of funds among the several armed services is often dictated as much by inter-service rivalry as by strategic or political decisions.

But despite all this, the military has not been able to sustain itself at its preferred size, to keep its strength constant or growing, or to retain for its use a fixed or growing portion of the Gross National Product. Even during the last two decades, the period of greatest military prominence, the size of the Army has varied enormously—from over 200 maneuver battalions in 1955, to 174 in 1965, rising to 217 at the peak of the Vietnam action in 1969, and then declining rapidly to 138 in 1972. Even military hardware, presumably of greater interest to the industrial side of the military-industrial complex, has often declined in quantity, even though per unit price has risen. The Navy had over 1,000 ships in 1955; it has only 700 today. The Air Force had nearly 24,000 aircraft in 1955; it has fewer than 14,000 today. This is not to say the combat strength of the military is substantially less than it once was, and there is greater firepower now at the disposal of each military unit, and there are various missile systems now in place, for which no earlier counterparts existed. But the total budget, and thus the total force level, of the military has been decided primarily by the President and not in any serious sense forced upon him by subordinates. (For example, President Truman decided to allocate one third of the federal budget to defense, President Eisenhower chose to spend no more than 10 per cent of the Gross National Product on it, and President Kennedy strongly supported Robert McNamara's radical and controversial budget revisions.) Even a matter of as great significance as the size of the total military budget for research and development has proved remarkably resistant to inflationary trends: In constant dollars, since 1964 that appropriation has been relatively steady (in 1972 dollars, about $30 billion a year).

The principal source of growth in the military budget in recent years has arisen from Congressionally-determined pay provisions. The legislature has voted for more or less automatic pay increases for military personnel with the result that the military budget has gone up even when the number of personnel in the military establishment has gone down.

The bureaucratic problems associated with the military establishment arise mostly from its internal management and are functions of its complexity, the uncertainty surrounding its future deployment, conflicts among its constituent services over mission and role, and the need to purchase expensive equipment without the benefit of a market economy that can control costs. Complexity, uncertainty, rivalry, and monopsony are inherent (and frustrating) aspects of the military as a bureaucracy, but they are very different problems from those typically associated with the phrase, "the military-industrial complex." The size and budget of the military are matters wholly within the power of civilian authorities to decide—indeed, the military budget contains the largest discretionary items in the entire federal budget.

If the Founding Fathers were to return to review their handiwork, they would no doubt be staggered by the size of both the Post Office and the Defense Department, and in the case of the latter, be worried about the implications of our commitments to various foreign powers. They surely would be amazed at the technological accomplishments but depressed by the cost and inefficiency of both departments; but they would not, I suspect, think that our Constitutional arrangements for managing these enterprises have proved defective or that there had occurred, as a result of the creation of these vast bureaus, an important shift in the locus of political authority.

They would observe that there have continued to operate strong localistic pressures in both systems—offices are operated, often uneconomically, in some small communities because small communities have influential Congressmen; military bases are maintained in many states because states have powerful Senators. But a national government with localistic biases is precisely the system they believed they had designed in 1787, and though they surely could not have then imagined the costs of it, they just as surely would have said (Hamilton possibly excepted) that these costs were the defects of the system's virtues.

Bureaucracy and clientelism

After 1861, the growth in the federal administrative system could no longer be explained primarily by an expansion of the postal service and other traditional bureaus. Though these continued to expand, new departments were added that reflected a new (or at least greater) emphasis on the enlargement of the scope of government.

Between 1861 and 1901, over 200,000 civilian employees were added to the federal service, only 52 per cent of whom were postal workers. Some of these, of course, staffed a larger military and naval establishment stimulated by the Civil War and the Spanish-American War. By 1901 there were over 44,000 civilian defense employees, mostly workers in government-owned arsenals and shipyards. But even these could account for less than one fourth of the increase in employment during the preceding 40 years.

What was striking about the period after 1861 was that the government began to give formal, bureaucratic recognition to the emergence of distinctive interests in a diversifying economy. As Richard L. Schott has written, "whereas earlier federal departments had been formed around specialized governmental functions (foreign affairs, war, finance, and the like), the new departments of this period—Agriculture, Labor, and Commerce—were devoted to the interests and aspirations of particular economic groups."

The original purpose behind these clientele-oriented departments was neither to subsidize nor to regulate, but to promote, chiefly by gathering and publishing statistics and (especially in the case of agriculture) by research. The formation of the Department of Agriculture in 1862 was to become a model, for better or worse, for later political campaigns for government recognition. A private association representing an interest—in this case the United States Agricultural Society—was formed. It made every President from Fillmore to Lincoln an honorary member, it enrolled key Congressmen, and it began to lobby for a new department. The precedent was followed by labor groups, especially the Knights of Labor, to secure creation in 1888 of a Department of Labor. It was broadened in 1903 to be a Department of Commerce and Labor, but 10 years later, at the insistence of the American Federation of Labor, the parts were separated and the two departments we now know were formed.

There was an early 19th-century precedent for the creation of these client-serving departments: the Pension Office, then in the Department of the Interior. Begun in 1833 and regularized in 1849, the Office became one of the largest bureaus of the government in the aftermath of the Civil War, as hundreds of thousands of Union Army veterans were made eligible for pensions if they had incurred a permanent disability or injury while on military duty; dependent widows were also eligible if their husbands had died in service or of service-connected injuries. The Grand Army of the Republic (GAR), the leading veterans' organization, was quick to exert pres-

sure for more generous pension laws and for more liberal administration of such laws as already existed. In 1879 Congressmen, noting the number of ex-servicemen living (and voting) in their states,
made veterans eligible for pensions retroactively to the date of their
discharge from the service, thus enabling thousands who had been
late in filing applications to be rewarded for their dilatoriness. In
1890 the law was changed again to make it unnecessary to have
been injured in the service—all that was necessary was to have
served and then to have acquired a permanent disability by any
means other than through "their own vicious habits." And whenever cases not qualifying under existing law came to the attention
of Congress, it promptly passed a special act making those persons
eligible by name.

So far as is known, the Pension Office was remarkably free of
corruption in the administration of this windfall—and why not, since
anything an administrator might deny, a legislator was only too
pleased to grant. By 1891 the Commissioner of Pensions observed
that his was "the largest executive bureau in the world." There were
over 6,000 officials supplemented by thousands of local physicians
paid on a fee basis. In 1900 alone, the Office had to process 477,000
cases. Fraud was rampant as thousands of persons brought false or
exaggerated claims; as Leonard D. White was later to write, "pen-
sioners and their attorneys seemed to have been engaged in a gi-
gantic conspiracy to defraud their own government." Though the
Office struggled to be honest, Congress was indifferent—or more accurately, complaisant: The GAR was a powerful electoral force and
it was ably and lucratively assisted by thousands of private pension
attorneys. The pattern of bureaucratic clientelism was set in a way
later to become a familiar feature of the governmental landscape—
a subsidy was initially provided, because it was either popular or
unnoticed, to a group that was powerfully benefited and had few
or disorganized opponents; the beneficiaries were organized to su-
pervise the administration and ensure the funding of the program;
the law authorizing the program, first passed because it seemed
the right thing to do, was left intact or even expanded because politically it became the only thing to do. A benefit once bestowed
cannot easily be withdrawn.

Public power and private interests

It was at the state level, however, that client-oriented bureaucracies proliferated in the 19th century. Chief among these were

the occupational licensing agencies. At the time of Independence, professions and occupations either could be freely entered (in which case the consumer had to judge the quality of service for himself) or entry was informally controlled by the existing members of the profession or occupation by personal tutelage and the management of reputations. The latter part of the 19th century, however, witnessed the increased use of law and bureaucracy to control entry into a line of work. The state courts generally allowed this on the grounds that it was a proper exercise of the "police power" of the state, but as Morton Keller has observed, "when state courts approved the licensing of barbers and blacksmiths, but not of horseshoers, it was evident that the principles governing certification were—to put it charitably—elusive ones." By 1952, there were more than 75 different occupations in the United States for which one needed a license to practice, and the awarding of these licenses was typically in the hands of persons already in the occupation, who could act under color of law. These licensing boards—for plumbers, dry cleaners, beauticians, attorneys, undertakers, and the like—frequently have been criticized as particularly flagrant examples of the excesses of a bureaucratic state. But the problems they create —of restricted entry, higher prices, and lengthy and complex initiation procedures—are not primarily the result of some bureaucratic pathology but of the possession of public power by persons who use it for private purposes. Or more accurately, they are the result of using public power in ways that benefited those in the profession in the sincere but unsubstantiated conviction that doing so would benefit the public generally.

The New Deal was perhaps the high water mark of at least the theory of bureaucratic clientelism. Not only did various sectors of society, notably agriculture, begin receiving massive subsidies, but the government proposed, through the National Industrial Recovery Act (NRA), to cloak with public power a vast number of industrial groupings and trade associations so that they might control production and prices in ways that would end the depression. The NRA's Blue Eagle fell before the Supreme Court—the wholesale delegation of public power to private interests was declared unconstitutional. But the piecemeal delegation was not, as the continued growth of specialized promotional agencies attests. The Civil Aeronautics Board, for example, erroneously thought to be exclusively a regulatory agency, was formed in 1938 "to promote" as well as to regulate civil aviation and it has done so by restricting entry and maintaining above-market rate fares.

Agriculture, of course, provides the leading case of clientelism. Theodore J. Lowi finds "at least 10 separate, autonomous, local self-governing systems" located in or closely associated with the Department of Agriculture that control to some significant degree the flow of billions of dollars in expenditures and loans. Local committees of farmers, private farm organizations, agency heads, and committee chairmen in Congress dominate policy-making in this area—not, perhaps, to the exclusion of the concerns of other publics, but certainly in ways not powerfully constrained by them.

"Cooperative federalism"

The growing edge of client-oriented bureaucracy can be found, however, not in government relations with private groups, but in the relations among governmental units. In dollar volume, the chief clients of federal domestic expenditures are state and local government agencies. To some degree, federal involvement in local affairs by the cooperative funding or management of local enterprises has always existed. The Northwest Ordinance of 1784 made public land available to finance local schools and the Morrill Act of 1862 gave land to support state colleges, but what Morton Grodzins and Daniel Elazar have called "cooperative federalism," though it always existed, did not begin in earnest until the passage in 1913 of the 16th Amendment to the Constitution allowed the federal government to levy an income tax on citizens and thereby to acquire access to vast sources of revenue. Between 1914 and 1917, federal aid to states and localities increased a thousandfold. By 1948 it amounted to over one tenth of all state and local spending; by 1970, to over one sixth.

The degree to which such grants, and the federal agencies that administer them, constrain or even direct state and local bureaucracies is a matter of dispute. No general answer can be given—federal support of welfare programs has left considerable discretion in the hands of the states over the size of benefits and some discretion over eligibility rules, whereas federal support of highway construction carries with it specific requirements as to design, safety, and (since 1968) environmental and social impact.

A few generalizations are possible, however. The first is that the states and not the cities have been from the first, and remain today, the principal client group for grants-in-aid. It was not until the Housing Act of 1937 that money was given in any substantial amount directly to local governments, and though many additional programs of this kind were later added, as late as 1970 less than 12 per cent of all federal aid went directly to cities and towns. The second general observation is that the 1960's mark a major watershed in the way in which the purposes of federal aid are determined. Before that time, most grants were for purposes initially defined by the states—to build highways and airports, to fund unemployment insurance programs, and the like. Beginning in the 1960's, the federal government, at the initiative of the President and his advisors, increasingly came to define the purposes of these grants—not necessarily over the objection of the states, but often without any initiative from them. Federal money was to be spent on poverty, ecology, planning, and other "national" goals for which, until the laws were passed, there were few, if any, well-organized and influential constituencies. Whereas federal money was once spent in response to the claims of distinct and organized clients, public or private, in the contemporary period federal money has increasingly been spent in ways that have *created* such clients.

And once rewarded or created, they are rarely penalized or abolished. What David Stockman has called the "social pork barrel" grows more or less steadily. Between 1950 and 1970, the number of farms declined from about 5.6 million to fewer than three million, but government payments to farmers rose from $283 million to $3.2 billion. In the public sector, even controversial programs have grown. Urban renewal programs have been sharply criticized, but federal support for the program rose from $281 million in 1965 to about $1 billion in 1972. Public housing has been enmeshed in controversy, but federal support for it rose from $206 million in 1965 to $845 million in 1972. Federal financial support for local poverty programs under the Office of Economic Opportunity has actually declined in recent years, but this cut is almost unique and it required the steadfast and deliberate attention of a determined President who was bitterly assailed both in the Congress and in the courts.

Self-perpetuating agencies

If the Founding Fathers were to return to examine bureaucratic clientelism, they would, I suspect, be deeply discouraged. James Madison clearly foresaw that American society would be "broken into many parts, interests and classes of citizens" and that this "multiplicity of interests" would help ensure against "the tyranny of the majority," especially in a federal regime with separate branches of government. Positive action would require a "coalition of a majority"; in the process of forming this coalition, the rights of all would be protected, not merely by self-interested bargains, but because in a free society such a coalition "could seldom take place on any other principles than those of justice and the general good." To those who wrongly believed that Madison thought of men as acting only out of base motives, the phrase is instructive: Persuading men who disagree to compromise their differences can rarely be achieved solely by the parceling out of relative advantage; the belief is also required that what is being agreed to is right, proper, and defensible before public opinion.

Most of the major new social programs of the United States, whether for the good of the few or the many, were initially adopted by broad coalitions appealing to general standards of justice or to conceptions of the public weal. This is certainly the case with most of the New Deal legislation—notably such programs as Social Security—and with most Great Society legislation—notably Medicare and aid to education; it was also conspicuously the case with respect to post-Great Society legislation pertaining to consumer and environmental concerns. State occupational licensing laws were supported by majorities interested in, among other things, the contribution of these statutes to public safety and health.

But when a program supplies particular benefits to an existing or newly-created interest, public or private, it creates a set of political relationships that make exceptionally difficult further alteration of that program by coalitions of the majority. What was created in the name of the common good is sustained in the name of the particular interest. Bureaucratic clientelism becomes self-perpetuating, in the absence of some crisis or scandal, because a single interest group to which the program matters greatly is highly motivated and well-situated to ward off the criticisms of other groups that have a broad but weak interest in the policy.

In short, a regime of separated powers makes it difficult to overcome objections and contrary interests sufficiently to permit the enactment of a new program or the creation of a new agency. Unless the legislation can be made to pass either with little notice or at a time of crisis or extraordinary majorities—and sometimes even then—the initiation of new programs requires public interest arguments. But the same regime works to protect agencies, once created, from unwelcome change because a major change is, in effect, new legislation that must overcome the same hurdles as the original law, but this time with one of the hurdles—the wishes of the agency and its client—raised much higher. As a result, the Madisonian system makes it relatively easy for the delegation of public power to private groups to go unchallenged and, therefore, for factional interests that have acquired a supportive public bureaucracy to rule without submitting their interests to the effective scrutiny and modification of other interests.

Bureaucracy and discretion

For many decades, the Supreme Court denied to the federal government any general "police power" over occupations and businesses, and thus most such regulation occurred at the state level and even there under the constraint that it must not violate the notion of "substantive due process"—that is, the view that there were sharp limits to the power of any government to take (and therefore to regulate) property. What clearly was within the regulatory province of the federal government was interstate commerce, and thus it is not surprising that the first major federal regulatory body should be the Interstate Commerce Commission (ICC), created in 1887.

What does cause, if not surprise, then at least dispute, is the view that the Commerce Act actually was intended to regulate railroads in the public interest. It has become fashionable of late to see this law as a device sought by the railroads to protect themselves from competition. The argument has been given its best-known formulation by Gabriel Kolko. Long-haul railroads, facing ruinous price wars and powerless to resist the demands of big shippers for rebates, tried to create voluntary cartels or "pools" that would keep rates high. These pools always collapsed, however, when one railroad or another would cut rates in order to get more business. To prevent this, the railroads turned to the federal government seeking a law to compel what persuasion could not induce. But the genesis of the act was in fact more complex: Shippers wanted protection from high prices charged by railroads that operated monopolistic services in certain communities; many other shippers served by competing lines wanted no legal barriers to prevent competition from driving prices down as far as possible; some railroads wanted regulation to ease competition, while others feared regulation. And the law as finally passed in fact made "pooling" (or cartels to keep prices up) illegal.

The true significance of the Commerce Act is not that it allowed public power to be used to make secure private wealth but that it created a federal commission with broadly delegated powers that would have to reconcile conflicting goals (the desire for higher or lower prices) in a political environment characterized by a struggle among organized interests and rapidly changing technology. In short, the Commerce Act brought forth a new dimension to the problem of bureaucracy: not those problems, as with the Post Office, that resulted from size and political constraints, but those that were caused by the need to make binding choices without any clear standards for choice.

The ICC was not, of course, the first federal agency with substantial discretionary powers over important matters. The Office of Indian Affairs, for a while in the War Department but after 1849 in the Interior Department, coped for the better part of a century with the Indian problem equipped with no clear policy, beset on all sides by passionate and opposing arguments, and infected with a level of fraud and corruption that seemed impossible to eliminate. There were many causes of the problem, but at root was the fact that the government was determined to control the Indians but could not decide toward what end that control should be exercised (extermination, relocation, and assimilation all had their advocates) and, to the extent the goal was assimilation, could find no method by which to achieve it. By the end of the century, a policy of relocation had been adopted *de facto* and the worst abuses of the Indian service had been eliminated—if not by administrative skill, then by the exhaustion of things in Indian possession worth stealing. By the turn of the century, the management of the Indian question had become the more or less routine administration of Indian schools and the allocation of reservation land among Indian claimants.

Regulation versus promotion

It was the ICC and agencies and commissions for which it was the precedent that became the principal example of federal discretionary authority. It is important, however, to be clear about just what this precedent was. Not everything we now call a regulatory agency was in fact intended to be one. The ICC, the Antitrust Division of the Justice Department, the Federal Trade Commission (FTC), the Food and Drug Administration (FDA), the National Labor Relations Board (NRLB)—all these *were* intended to be genuinely regulatory bodies created to handle under public auspices matters once left to private arrangements. The techniques they were to employ varied: approving rates (ICC), issuing cease-

and-desist orders (FTC), bringing civil or criminal actions in the courts (the Antitrust Division), defining after a hearing an appropriate standard of conduct (NRLB), or testing a product for safety (FDA). In each case, however, Congress clearly intended that the agency either define its own standards (a safe drug, a conspiracy in restraint of trade, a fair labor practice) or choose among competing claims (a higher or lower rate for shipping grain).

Other agencies often grouped with these regulatory bodies—the Civil Aeronautics Board, the Federal Communications Commission, the Maritime Commission—were designed, however, not primarily to regulate, but to *promote* the development of various infant or threatened industries. However, unlike fostering agriculture or commerce, fostering civil aviation or radio broadcasting was thought to require limiting entry (to prevent "unsafe" aviation or broadcast interference); but at the time these laws were passed few believed that the restrictions on entry would be many, or that the choices would be made on any but technical or otherwise noncontroversial criteria. We smile now at their naïveté, but we continue to share it —today we sometimes suppose that choosing an approved exhaust emission control system or a water pollution control system can be done on the basis of technical criteria and without affecting production and employment.

Majoritarian politics

The creation of regulatory bureaucracies has occurred, as is often remarked, in waves. The first was the period between 1887 and 1890 (the Commerce Act and the Antitrust Act), the second between 1906 and 1915 (the Pure Food and Drug Act, the Meat Inspection Act, the Federal Trade Commission Act, the Clayton Act), the third during the 1930's (the Food, Drug, and Cosmetic Act, the Public Utility Holding Company Act, the Securities Exchange Act, the Natural Gas Act, the National Labor Relations Act), and the fourth during the latter part of the 1960's (the Water Quality Act, the Truth in Lending Act, the National Traffic and Motor Vehicle Safety Act, various amendments to the drug laws, the Motor Vehicle Pollution Control Act, and many others).

Each of these periods was characterized by progressive or liberal Presidents in office (Cleveland, T. R. Roosevelt, Wilson, F. D. Roosevelt, Johnson); one was a period of national crisis (the 1930's); three were periods when the President enjoyed extraordinary majorities of his own party in both houses of Congress (1914-

1916, 1932-1940, and 1964-1968); and only the first period preceded the emergence of the national mass media of communication. These facts are important because of the special difficulty of passing any genuinely regulatory legislation: A single interest, the regulated party, sees itself seriously threatened by a law proposed by a policy entrepreneur who must appeal to an unorganized majority, the members of which may not expect to be substantially or directly benefitted by the law. Without special political circumstances—a crisis, a scandal, extraordinary majorities, an especially vigorous President, the support of media—the normal barriers to legislative innovation (i.e., to the formation of a "coalition of the majority") may prove insuperable.

Stated another way, the initiation of regulatory programs tends to take the form of majoritarian rather than coalitional politics. The Madisonian system is placed in temporary suspense: Exceptional majorities propelled by a public mood and led by a skillful policy entrepreneur take action that might not be possible under ordinary circumstances (closely divided parties, legislative-executive checks and balances, popular indifference). The consequence of majoritarian politics for the administration of regulatory bureaucracies is great. To initiate and sustain the necessary legislative mood, strong, moralistic, and sometimes ideological appeals are necessary—leading, in turn, to the granting of broad mandates of power to the new agency (a modest delegation of authority would obviously be inadequate if the problem to be resolved is of crisis proportions), or to the specifying of exacting standards to be enforced (e.g., no carcinogenic products may be sold, 95 per cent of the pollutants must be elminated), or to both.

Either in applying a vague but broad rule ("the public interest, convenience, and necessity") or in enforcing a clear and strict standard, the regulatory agency will tend to broaden the range and domain of its authority, to lag behind technological and economic change, to resist deregulation, to stimulate corruption, and to contribute to the bureaucratization of private institutions.

It will broaden its regulatory reach out of a variety of motives: to satisfy the demand of the regulated enterprise that it be protected from competition, to make effective the initial regulatory action by attending to the unanticipated side effects of that action, to discover or stretch the meaning of vague statutory language, or to respond to new constituencies induced by the existence of the agency to convert what were once private demands into public pressures. For example, the Civil Aeronautics Board, out of a desire

both to promote aviation and to protect the regulated price structure of the industry, will resist the entry into the industry of new
carriers. If a Public Utilities Commission sets rates too low for a
certain class of customers, the utility will allow service to those
customers to decline in quality, leading in turn to a demand that
the Commission also regulate the quality of service. If the Federal
Communications Commission cannot decide who snould receive a
broadcast license by applying the "public interest" standard, it will
be powerfully tempted to invest that phrase with whatever preferences the majority of the Commission then entertains, leading in
turn to the exercise of control over many more aspects of broadcasting than merely signal interference—all in the name of deciding
what the standard for entry shall be. If the Antitrust Division can
prosecute conspiracies in restraint of trade, it will attract to itself
the complaints of various firms about business practices that are
neither conspiratorial nor restraining but merely competitive, and a
"vigorous" antitrust lawyer may conclude that these practices warrant prosecution.

Bureaucratic inertia

Regulatory agencies are slow to respond to change for the same
reason all organizations with an assured existence are slow: There
is no incentive to respond. Furthermore, the requirements of due
process and of political conciliation will make any response time
consuming. For example, owing to the complexity of the matter and
the money at stake, any comprehensive review of the long-distance
rates of the telephone company will take years, and possibly may
take decades.

Deregulation, when warranted by changed economic circumstances or undesired regulatory results, will be resisted. Any organization, and *a fortiori* any public organization, develops a genuine
belief in the rightness of its mission that is expressed as a commitment to regulation as a process. This happened to the ICC in the
early decades of this century as it steadily sought both enlarged
powers (setting minimum as well as maximum rates) and a broader jurisdiction (over trucks, barges, and pipelines as well as railroads). It even urged incorporation into the Transportation Act of
1920 language directing it to prepare a comprehensive transportation plan for the nation. Furthermore, any regulatory agency will
confer benefits on some group or interest, whether intended or not;
those beneficiaries will stoutly resist deregulation. (But in happy

proof of the fact that there are no iron laws, even about bureaucracies, we note the recent proposals emanating from the Federal
Power Commission that the price of natural gas be substantially
deregulated.)

The operation of regulatory bureaus may tend to bureaucratize
the private sector. The costs of conforming to many regulations can
be met most easily—often, *only*—by large firms and institutions with
specialized bureaucracies of their own. Smaller firms and groups
often must choose between unacceptably high overhead costs, vio-
lating the law, or going out of business. A small bakery producing
limited runs of a high-quality product literally may not be able to
meet the safety and health standards for equipment, or to keep
track of and administer fairly its obligations to its two employees;
but unless the bakery is willing to break the law, it must sell out to
a big bakery that can afford to do these things, but may not be
inclined to make and sell good bread. I am not aware of any data
that measure private bureaucratization or industrial concentration
as a function of the economies of scale produced by the need to
cope with the regulatory environment, but I see no reason why
such data could not be found.

Finally, regulatory agencies that control entry, fix prices, or substantially affect the profitability of an industry create a powerful
stimulus for direct or indirect forms of corruption. The revelations
about campaign finance in the 1972 presidential election show dramatically that there will be a response to that stimulus. Many corporations, disproportionately those in regulated industries (airlines,
milk producers, oil companies), made illegal or hard to justify campaign contributions involving very large sums.

The era of contract

It is far from clear what the Founding Fathers would have
thought of all this. They were not doctrinaire exponents of laissez
faire, nor were 18th-century governments timid about asserting
their powers over the economy. Every imaginable device of fiscal
policy was employed by the states after the Revolutionary War.
Mother England had, during the mercantilist era, fixed prices and
wages, licensed merchants, and granted monopolies and subsidies.
(What were the royal grants of American land to immigrant settlers but the greatest of subsidies, sometimes—as in Pennsylvania—
almost monopolistically given?) European nations regularly operated state enterprises, controlled trade, and protected industry. But

as William D. Grampp has noted, at the Constitutional Convention the Founders considered authorizing only four kinds of economic controls, and they rejected two of them. They agreed to allow the Congress to regulate international and interstate commerce and to give monopoly protection in the form of copyrights and patents. Even Madison's proposal to allow the federal government to charter corporations was rejected. Not one of the 85 *Federalist* papers dealt with economic regulation; indeed, the only reference to commerce was the value to it of a unified nation and a strong navy.

G. Warren Nutter has speculated as to why our Founders were so restrained in equipping the new government with explicit regulatory powers. One reason may have been the impact of Adam Smith's *Wealth of Nations,* published the same year as the Declaration of Independence, and certainly soon familiar to many rebel leaders, notably Hamilton. Smith himself sought to explain the American prosperity before the Revolution by the fact that Britain, through "salutary neglect," had not imposed mercantilist rules on the colonial economy. "Plenty of good land, and liberty to manage their own affairs in their own way" were the "two great causes" of colonial prosperity. As Nutter observes, there was a spirit of individualistic venture among the colonies that found economic expression in the belief that voluntary contracts were the proper organization principle of enterprise.

One consequence of this view was that the courts in many states were heavily burdened with cases testing the provisions of contracts and settling debts under them. In one rural county in Massachusetts the judges heard over 800 civil cases during 1785. As James Willard Hurst has written, the years before 1875 were "above all else, the years of contract in our law."

The era of contract came to an end with the rise of economic organizations so large or with consequences so great that contracts were no longer adequate, in the public's view, to adjust corporate behavior to the legitimate expectations of other parties. The courts were slower to accede to this change than were many legislatures, but in time they acceded completely, and the era of administrative regulation was upon us. The Founders, were they to return, would understand the change in the scale and social significance of enterprise, would approve of many of the purposes of regulation, perhaps would approve of the behavior of some of the regulatory bureaus seeking to realize those purposes, but surely would be dismayed at the political cost resulting from having vested vast discretionary authority in the hands of officials whose very existence

—to say nothing of whose function—was not anticipated by the Constitutional Convention, and whose effective control is beyond the capacity of the governing institutions which that Convention had designed.

The bureaucratic state and the Revolution

The American Revolution was not only a struggle for independence but a fundamental rethinking of the nature of political authority. Indeed, until that reformulation was completed the Revolution was not finished. What made political authority problematic for the colonists was the extent to which they believed Mother England had subverted their liberties despite the protection of the British constitution, until then widely regarded in America as the most perfect set of governing arrangements yet devised. The evidence of usurpation is now familiar: unjust taxation, the weakening of the independence of the judiciary, the stationing of standing armies, and the extensive use of royal patronage to reward office-seekers at colonial expense. Except for the issue of taxation, which raised for the colonists major questions of representation, almost all of their complaints involved the abuse of *administrative* powers.

The first solution proposed by Americans to remedy this abuse was the vesting of most (or, in the case of Pennsylvania and a few other states, virtually all) powers in the legislature. But the events after 1776 in many colonies, notably Pennsylvania, convinced the most thoughtful citizens that legislative abuses were as likely as administrative ones: In the extreme case, citizens would suffer from the "tyranny of the majority." Their solution to this problem was, of course, the theory of the separation of powers by which, as brilliantly argued in *The Federalist* papers, each branch of government would check the likely usurpations of the other.

This formulation went essentially unchallenged in theory and unmodified by practice for over a century. Though a sizeable administrative apparatus had come into being by the end of the 19th century, it constituted no serious threat to the existing distribution of political power because it either performed routine tasks (the Post Office) or dealt with temporary crises (the military). Some agencies wielding discretionary authority existed, but they either dealt with groups whose liberties were not of much concern (the Indian Office) or their exercise of discretion was minutely scrutinized by Congress (the Land Office, the Pension Office, the Customs Office). The major discretionary agencies of the 19th century flourished at the very period of greatest Congressional domination of the political process—the decades after the Civil War—and thus, though their supervision was typically inefficient and sometimes corrupt, these agencies were for most practical purposes direct dependencies of Congress. In short, their existence did not call into question the theory of the separation of powers.

But with the growth of client-serving and regulatory agencies, grave questions began to be raised—usually implicitly—about that theory. A client-serving bureau, because of its relations with some source of private power, could become partially independent of both the executive and legislative branches—or in the case of the latter, dependent upon certain committees and independent of others and of the views of the Congress as a whole. A regulatory agency (that is to say, a truly regulatory one and not a clientelist or promotional agency hiding behind a regulatory fig leaf) was, in the typical case, placed formally outside the existing branches of government. Indeed, they were called "independent" or "quasi-judicial" agencies (they might as well have been called "quasi-executive" or "quasi-legislative") and thus the special status that clientelist bureaus achieved *de facto*, the regulatory ones achieved *de jure*.

It is, of course, inadequate and misleading to criticize these agencies, as has often been done, merely because they raise questions about the problem of sovereignty. The crucial test of their value is their behavior, and that can be judged only by applying economic and welfare criteria to the policies they produce. But if such judgments should prove damning, as increasingly has been the case, then the problem of finding the authority with which to alter or abolish such organizations becomes acute. In this regard the theory of the separation of powers has proved unhelpful.

The separation of powers makes difficult, in ordinary times, the extension of public power over private conduct—as a nation, we came more slowly to the welfare state than almost any European nation, and we still engage in less central planning and operate fewer nationalized industries than other democratic regimes. But we have extended the regulatory sway of our national government as far or farther than that of most other liberal regimes (our environmental and safety codes are now models for much of Europe), and the bureaus wielding these discretionary powers are, once created, harder to change or redirect than would be the case if authority were more centralized.

The shift of power toward the bureaucracy was not inevitable.

It did not result simply from increased specialization, the growth of industry, or the imperialistic designs of the bureaus themselves. Before the second decade of this century, there was no federal bureaucracy wielding substantial discretionary powers. That we have one now is the result of political decisions made by elected representatives. Fifty years ago, the people often wanted more of government than it was willing to provide—it was, in that sense, a republican government in which representatives moderated popular demands. Today, not only does political action follow quickly upon the stimulus of public interest, but government itself creates that stimulus and sometimes acts in advance of it.

All democratic regimes tend to shift resources from the private to the public sector and to enlarge the size of the administrative component of government. The particularistic and localistic nature of American democracy has created a particularistic and client-serving administration. If our bureaucracy often serves special interests and is subject to no central direction, it is because our legislature often serves special interests and is subject to no central leadership. For Congress to complain of what it has created and it maintains is, to be charitable, misleading. Congress could change what it has devised, but there is little reason to suppose it will.

5

Towards
an
imperial
judiciary?

NATHAN GLAZER

. . . The justices wear black gowns, being not merely the only public officers, but the only non-ecclesiastical persons of any kind whatever within the bounds of the United States who use any official dress.
—The American Commonwealth

NON-LAWYER who considers the remarkable role of courts in the interpretation of the Constitution and the laws in the United States finds himself in a never-never land—one in which questions he never dreamed of raising are discussed at incredible length, while questions that would appear to be the first to come to mind are hardly ever raised. This is particularly the case when the concern of the non-professional observer is with social policy rather than with constitutional law as such.

Thus, fine distinctions in the use of evidence in criminal cases are debated at length, and used as a basic test in judging whether a court or individual justices are liberal or conservative; whereas one sees little discussion of why judges may hold up hundreds of millions of dollars in federal payments to states and cities to force them to make numerous civil service appointments, or why judges may require a state or city to provide extensive and specifically defined social services. The original *Brown* decision on school desegregation was properly debated at length by our leading constitutional authorities. But what are arguably the most disruptive decisions ever made by the courts—the requirement that children be bused to distant schools—have excited much less professional interest than popular interest.

One reason for this disparity is that judges and lawyers are trained to see continuity in the development of constitutional law. However far-reaching the actions the courts take, the lawyers who propose such actions and the judges who rule on them are committed by the logic of legal reasoning to insist that they are only unveiling a rule that existed all along in the recesses of the Constitution or the bowels of legislation: Nothing new has been added, they say, even though great consequences follow from their decisions.

Political scientists who study the courts are somewhat freer to see truly original developments in the constitutional law, but they generally do not go beyond interpreting these developments as part of a cycle. If the court changes, to them it changes within a well-understood pattern, in which periods of activism symmetrically contrast with periods of quietism: At some point the Supreme Court, exercising its power to interpret the Constitution and the laws and to overrule the interpretations of legislatures and executives,[1] goes too far—thus, we have the *Dred Scott* decision, or the decisions overruling the actions of the New Deal in the mid-1930's. An explosion then results: as a result of the *Dred Scott* decision, a war; as a result of the anti-New Deal decisions, the court-packing plan. To the political scientist, the court follows not only its own logic but the logic of public opinion. Since it is without the independent power to enforce its decrees, the Court then withdraws. Its withdrawal is assisted ultimately by the appointment power of the President, who is in closer touch with public opinion. A period of quietism thus succeeds a period of activism. This is a reasonable description of the pattern of judicial interpretation, supported by history and by American constitutional arrangements, and this is as far as one of our best-known analysts of the Supreme Court, Robert McCloskey, went.

There is, however, another perspective on judicial interpretation, one which is generally identified with outraged legislators, Presidents, governors, and people: The Court has gone too far, they say,

---

[1] In this article, I do not plan to go into the question of the sources of this power, whether given in the Constitution, or seized by Chief Justice Marshall, or properly or improperly established—it is there, and it is permanent.

and the return of the quietistic phase of the cycle will not satisfy us, for the Court is engaged in a damaging and unconstitutional revolution that even the cyclical return of a period of quietism cannot curb. The lawyers who must operate within the assumption of continuity are inclined to dismiss such outrage, and the legal commentators who look at the long stretch of history are sure that quietism will replace activism—as it has before—and that the courts will retire from the front pages of the newspapers. Yet in 1975, all the evidence suggests that this third perspective is really the correct one: The courts truly have changed their role in American life. American courts, the most powerful in the world—they were that already when Tocqueville wrote and when Bryce wrote—are now far more powerful than ever before; public opinion—which Tocqueville, Bryce, and other analysts thought would control the courts as well as so much else in American life—is weaker. The legislatures and the executive now moderate their outbursts, for apparently outbursts will do no good. And courts, through interpretation of the Constitution and the laws, now reach into the lives of the people, against the will of the people, deeper than they ever have in American history.

From Warren to Burger

These are sweeping assertions, yet the course of the law since 1969 supports them. For in 1969 something was supposed to happen, and didn't. In his first term, President Nixon, who opposed the Warren Court's activism, succeeded in making four appointments, and the period of activism that had begun with the *Brown* decision of 1954 was supposed to come to an end, as the Warren Court was replaced by the Burger Court. Instead, there have been more far-reaching decisions—if estimated by the impact on people and their everyday lives—since 1969 than in 1954-1968, even with four Nixon-appointed Justices. In 1971 the *Swann* decision for the first time legitimated massive busing of children to overcome segregation in a large city; in 1974 the *Keyes* decision for the first time legitimated such massive busing in a Northern city, and in addition legitimated standards of proof for *de jure* segregation that were so loose that it guaranteed that *de jure* segregation could be found everywhere (which meant that the Court's narrow 5-4 decision in *Milliken*—which overturned lower court requirements for the merger of Detroit and its suburbs in order to create, through busing, schools with smaller proportions of black children—could very likely be circumvented by demonstrating that the suburbs, too, were engaged in *de jure* segregation). In 1971 the Court legitimated federal government guidelines for the use of tests in employment that required strict standards of "job-relatedness" if, on the basis of such tests, differing proportions of certain ethnic and racial groups were given employment. This decision has, in effect, declared illegal most efforts by employers, public and private, to hire more qualified employees. Since apparently all tests select differing proportions of one group or another, and few tests can be shown to be job-related by the strict standards of the guidelines, most employee hiring on the basis of tests now can be labeled discriminatory and employers may be required, either by lower court orders or by Department of Justice consent decrees, to hire by racial and ethnic quotas—a practice which is specifically forbidden by the Civil Rights Act of 1964 and (one would think) unconstitutional because of its denial of the "equal protection of the laws." In 1973, the Burger Court ruled in *Roe* and *Doe* that just about all state laws on abortion were unconstitutional and decreed that state laws must treat each third of the pregnancy period according to different standards. In 1975 it spread the awesome limitations of "due process" to the public schools, which now could not restrict the constitutional rights of students by suspending or expelling them without at least something resembling a criminal trial. In 1975 it agreed, in the first of a series of important cases on the rights of mental patients, that harmless persons could not be detained involuntarily in mental hospitals.

The list could be extended. It is true that the Court stayed its hand in other cases that could have had enormous consequences: In particular, it refused to accept the argument that states must ensure that financial support of schools be unaffected by differences in community wealth, and it did not allow a challenge by inner-city residents to suburban zoning ordinances. But in both areas state courts are active, and it is hard to see what is to prevent them from decreeing for their states (as the courts of New Jersey already have) the revolution in equalization of school financing and zoning that the Supreme Court has refused to decree for the nation.

What was most striking is that all of these cases—and many others extending the reach of government, whether it wished it or not, into the lives of people, and of the courts over the actions of legislatures and communities—were made despite the rapid appointment of four Justices by President Nixon. The Nixon Justices either supported the majority in 9-0 decisions (e.g., *Swann*), or split (thus only one of the four dissented from the majority in *Doe* and *Roe*), and only rarely (in *Goss,* on due process for students) voted as a bloc against the majority of five who had served on the Warren Court.

Supreme Court analysts and reporters very often tell a different story from this, because in judging the Court, they tend to focus on the endless details of what is or is not allowed in criminal law—the use of confessions, searches, and the like. Emphasizing the criminal law, they apparently see a retreat where another observer might see a modicum of common sense. But I believe all agree that the Burger Court has been surprisingly like the Warren Court; and this raises the question of why the expected turn has not yet taken place—seven years after Nixon was elected, and after four of his appointees were on the court.

The end of a conservative judiciary?

Three characteristics of the Burger Court, and of the Warren Court, have excited less interest than they might have, and would suggest that we must at least consider the possibility that there has been a permanent change in the character of the courts and their role in the commonwealth, rather than simply a somewhat extended activist cycle.

First, the activist cycles of the past have always been characterized by conservative Courts, acting to restrict liberal Congresses, Presidents, or state governments—the Marshall Court, the Taney Court, and the Taft-Hughes Courts. This made sense: The Court, after all, was designed by the Founders to be a conservative institution, a check on popularly elected legislatures and an elected (even if not at the beginning *popularly* elected) President. The Court was appointed, and it held tenure for life. It was, as Alexander Bickel wrote, paraphrasing Hamilton in the 78th *Federalist* paper, "the least dangerous branch." To quote Hamilton: "Whoever attentively considers the different departments of power must perceive, that, in a government in which they are separated from each other, the judiciary, from the nature of its functions, will always be the least dangerous to the political rights of the Constitution; because it will be least in capacity to annoy or injure them. The Executive not only dispenses the honors, but holds the sword of the community; the legislature not only commands the purse, but prescribes the rules by which the duties and rights of every citizen are to be regulated." (As we shall see, however, the legislature no longer controls the purse, if the Court rules otherwise, nor prescribes the rules governing duties and rights—though the sword is still in the hands of the Executive.)

But something extraordinary has happened when a liberal Court confronts a conservative executive and legislature, as the Warren Court did after the election of President Eisenhower: The *natural* expectations, the order of history, have been reversed. It is even more extraordinary when, after 15 years of appointments to the Supreme Court and the subordinate courts by conservative Presidents (against seven years by liberal Presidents), this strange posture still persists.

A second extraordinary feature of the post-1954 activism is a corollary in part of the first: In the past the role of activist courts was to *restrict* the executive and legislature in what they could do. The distinctive characteristic of more recent activist courts has been to *extend* the role of what the government could do, even when the government did not want to do it. The *Swann* and *Keyes* decisions meant that government *must* move children around to distant schools against the will of their parents. The *Griggs* decision meant that government *must* monitor the race and ethnicity of job applicants and test-takers. The cases concerning the rights of mental patients and prisoners, which are for the most part still in the lower courts, say that government *must* provide treatment and rehabilitation whether it knows how or not. Federal Judge Weinstein's ruling in a New York school desegregation case seems to say that government *must* racially balance communities. And so on.

An interesting example of this unwilled extension of governmental action is that of the Environmental Protection Agency (EPA). It did not wish to issue rules preserving pure air in areas without pollution or imposing drastic transportation controls. To the EPA, this did not seem to be what Congress intended; but under court order, it was required to do both. Similarly, the Department of Health, Education, and Welfare (HEW) apparently did not want to move against the Negro colleges of the South, now no longer segregated under law but still with predominantly black enrollments, nor was this in the interests of those colleges, or their students, or indeed anyone else—but Federal judges required HEW to do so.

In these, as in other cases, government is required to do what the Congress did not order it to do and may well oppose, what the executive does not feel it wise to do, and most important what it does not know how to do. How *does* one create that permanently racially balanced community that Judge Weinstein wants so that the schools may be permanently racially balanced? How does one

create that good community in Boston public housing that Judge
Garrity wants so that vandalism repair costs may be brought down
to what the authority can afford? How does one rehabilitate prisoners? Or treat mental patients? Like Canute, the Judges decree the
sea must not advance, and weary administrators—hectored by enthusiastic, if ignorant, lawyers for public advocacy centers—must go
through the motions to show the courts they are trying.

Reconsidering the cyclical theory

A third feature of the new activism which is also extraordinary:
The Court's actions now seem to arouse fewer angry reactions from
the people and the legislatures. The power of the Court has been
exercised so often and so successfully over the last 20 years, and the
ability to restrict or control it by either new legislation, constitutional amendment, or new appointments has met with such uniform failure, that the Court, and the subordinate courts, are now
seen as forces of nature, difficult to predict and impossible to control. Thus, one may contrast the outburst over the school prayer
decisions of 1962-3 with the relative quietude of response to the
abortion decisions of 1973. Or, contrast the effort to adopt a constitutional amendment to control the sweep of reapportionment
decisions in 1964-66 with the general view in Congress today that
any effort to control the Court on busing by means of a constitutional amendment has no chance of succeeding.

This is, of course, not necessarily witness to the strength of the
Court as such: What it reflects, in addition, is the agreement of
large sectors of opinion—even if it is still minority opinion—with the
Court's actions. But this opinion in favor of the Court is shaped by
the reserves of strength the Court possesses: the positive opinion of
the Court in the dominant mass media—the national television news
shows, the national news magazines, and the most influential newspapers; and the bias in its favor among the informed electorate
generally, and among significant groups of opinion-leaders. The
Court is further the beneficiary of two accidents of political history
(or at least they may be accidents): Because he was running
against Barry Goldwater, Lyndon Johnson had an overwhelming
victory in 1964, and helped bring into office so many liberal Congressmen that the powerful effort to limit the Court on reapportionment was blunted. And because of the Watergate scandal, so many
liberal Congressmen were returned to Congress in 1974 that the
natural life of the activist cycle in the Court's history was extended

at least two years, and perhaps longer. Thus this Congress will not start any amendment process to limit the Court on busing. (It is understood that liberal Congressmen today, as against our previous history, support the power of the Court. If new appointments bring about the expected conservative switch, that position may change.)

The key point, however, is that the major limitation on the Court's power—public opinion, expressing itself through the Presidency and the Congress—has not come into effect to limit this Court. Outrage at its actions was stronger 12 or 13 years ago than it is today, though the intrusive reach of the Court's actions into the daily life of citizens has become much stronger. Ironically, the President who once wanted to impeach Justice Douglas may well find that the Justice who has served longest on the Supreme Court may survive his own Presidential term.

What I am suggesting is that we must reconsider the theory that activist cycles are succeeded by quiescent ones. This belief was based on the view that public opinion in the end controls the Court, which has the power of neither purse nor sword, and that the Court is thus still pretty much where the Founders and Chief Justice Marshall established it, as one of the three coequal branches of government, with great moral authority but little else. In contrast, it appears that the controls on Court power have become obsolescent and that the role of the Court—and courts generally—has changed significantly, such that the most powerful Court and Judiciary in the world have become even more powerful, raising questions of some gravity for the Commonwealth. Of course, any long-range view shortly may be made irrelevant by current events. Two more conservative appointments, one might think, and the Court will revert, only eight years later than one might have expected, to its quietistic phase, and President and Congress will resume their positions of dominance. But there are a number of other factors, which must at least be considered, which would argue that the matter is more serious.

The factors affecting the Court's power transcend, I believe, the question of the individual outlooks and philosophies of the present Justices or their potential successors. It is true that in an institution in which individuality is so dominant that 5-4 decisions on vital matters affecting the nation are common—with no apparent influence that can be exerted on the minority to change its vote so that the nation may accept these decisions with better grace—the character of individual Justices is not a matter to be taken lightly. However, there are three factors that argue that the activist phase of the present Court will not easily be reversed, and two of them are quite new.

The new Constitutional logic

The first factor is well-known and broadly discussed in the constitutional law literature: The Court must work out the logic of positions once taken, and it cannot easily withdraw from the implications of these positions. Thus, if "standing" to sue has been radically expanded so that many interests and individuals who in the past had no access to constitutional adjudication of their claims now have such access, a systemic change has occurred, and it is not possible to revert to an earlier, more restrictive view of "standing."

Whether one calls it a constitutional revolution or not—and such excellent analysts as Philip Kurland argue that there was substantial continuity by the Warren Court with past rulings—the power to enter litigation has been greatly expanded in recent years, and new rulings have been laid down on substantive issues which ensure that a great deal of new litigation will ensue to establish their bounds. Archibald Cox, writing from the perspective of his service as Solicitor General under Presidents Kennedy and Johnson—long before some of the most radical decisions were made, in the post-1971 Court—made this revealing summary of the situation in 1967:

> The Warren Court has been quick to slough off the restraints its predecessors erected for deciding whether and when the Court will adjudicate constitutional issues. The precepts of wise constitutional adjudication that were taught law students in the 1930's counseled postponement and avoidance . . . interference with the work of a coordinate branch of the federal government or a sovereign state was thought justifiable only in cases of absolute necessity. The party attacking the statute had to show its unconstitutionality applied to him. . . . The plaintiff must demonstrate his standing by showing that he was injured. . . . Equity would not enjoin the enforcement of an allegedly unconstitutional statute where the unconstitutionality could be raised as a defense in a criminal prosecution, except in the most exceptional conditions. The constitutional battles of the 1930's were often fought on these procedural grounds. Those seeking to sustain new governmental activities . . . could rest satisfied with blocking judicial intervention. . . .
>
> Today, the attitude is changed, and the rules of judicial self-restraint that looked to the avoidance of constitutional rulings have been eroded in opinions strongly suggesting that the present Court feels it has a responsibility to make its influence felt in support or check of other branches of government, or in innovation, even though not coerced by the necessities of litigation.

There is still some life in the old restraints, as we may see in the zoning case, where the Court in a 5-4 decision did not accept that inner-city residents in general were damaged by suburban zoning practices. But a Court cannot radically and totally ignore precedent, particularly when it has created expectations. It is still a court of law. And so, the logic of decisions once taken must be worked out. If due process is now something that must be taken into account not only by the legislatures and executives, but by schools and colleges and businesses, the logic of that position must be worked out. A line undoubtedly will be set at some point, but where that will be is uncertain.

Of course, it must be understood, it is not only the Court that expands the area of litigation through new rulings. Just as the executive is required by the Court to institute procedures and rulings and rights it would prefer not to (as in the case of the EPA), the establishment of new ground for legal action is also the result of legislative and executive action, and the courts must enforce these new laws. Thus the reach of due process is extended not only by the courts—though I believe they have played the largest roles—but by legislatures. The Colorado legislature has recently required due process in all cases of dismissal or non-renewal of appointment in many of the state's colleges—an act which has been described as instituting "instant tenure." Due process is even more significantly extended by executive action implementing legislation, as in the recent HEW guidelines implementing non-discrimination by sex in federally-supported education activities. As is known, these guidelines ban single-sex physical education classes (with certain exceptions), and require each school to set up grievance procedures for complaints of discrimination on grounds of sex. Many Congressmen thought this went beyond the legislation Congress had passed, but in 1975 there was a different Congress from that which passed the Higher Education Act in 1972—perhaps an aberrant Congress as a result of Watergate—and this Congress, it seems, will not intervene. In any case, once the elaborate process of legislation has worked its course, it is hard for Congress, owing to its clumsy and complex procedures, to control implementation of legislation by the executive, and it is impossible for Congress to control the interpretation by the judiciary of that implementation.

However, even if due process expands as a result not only of the actions of courts but also of legislatures and executives, it does so on the basis of the teaching of the Court, a teaching almost universally applauded by those who are considered qualified to judge. If the Court expands due process in every sphere, and teaches that this is the teaching of the Constitution, then it is no surprise if legislatures and executives follow that teaching in good measure on their own.

Broadening the reach of legal principles

The logic of the wider expansion of "standing" and of due process and of equal treatment of the laws, we may be sure, has not yet been worked out: There is a good deal more to come. "More to come" means a continued and powerfully intrusive role for the courts that they cannot avoid. We have gone so far in just a few years that cautions uttered just a few years ago now seem archaic. Thus, eight years ago, Archibald Cox wrote with some discomfort on the expansion of the doctrine that private action came under constitutional protection if it involved a "state interest":

> How could a court rule that discrimination at lunch counters violated the 14th Amendment without going on to rule that the amendment is also violated by discrimination in employment, in admitting pupils to private colleges, and in the sale and rental of housing? It would immeasurably advance the cause of human justice to have on the statute books open housing laws, fair employment practices acts, and the like; but it would be amazing to find that all the hard legislative fights were unnecessary and the will of the people is irrelevant because the legal requirements were long ago written into the 14th Amendment. . . .
>
> Perhaps the consequence [of the expansion of the notion of state interest] should not frighten us . . . but I wonder whether we should not pay a heavy price in terms of the loss to the richness, variety, and the initiative of our present pluralistic society. . . .
>
> Consider the effect upon educational institutions. Though few will defend racial discrimination, one can fairly ask whether Notre Dame should be barred from preferring Roman Catholics or Baylor from giving preference to Baptists. . . . Should [every school's or college's] admission practices and perhaps its examinations be subject to judicial scrutiny for due process of law?

Clearly all this, in large measure, has come to pass. A recent decision of the respected Second Circuit has decreed that actions by foundations also require due process—after all, foundations are set up under state law, are granted tax exemption, are regulated by legislation, etc. And so every disappointed grant-seeker will have access to a due process whose full dimensions will have to be determined by other courts and other cases.

The working out of the logic of a position is not only the work of the Supreme Court. It is also the work of the lower federal courts, the circuit courts, the state courts. They are also taught by the Supreme Court. As the right to sue expands, as the meaning of due process and equal protection is broadened, more and more kinds of actions in the courts to expand the reach of social policy become possible.

Most cases, of course, never reach the Supreme Court. The law is then established under the lower courts and an egregious and unchallenged intrusiveness of the courts spreads under the general protection given by some larger decisions. Conceivably, the Supreme Court itself might find objectionable the actions of some lower court judge to implement its decisions. The Court, for example, might object to a lottery, decreed for school assignments by a Charlotte judge, which ties students to schools on the basis of race, regardless of their changes in residence; or to the banning of anti-busing meetings by a Denver judge; or to the ethnic quotas decreed for the selective secondary schools in Boston by a Boston judge; or to the same judge's wholesale recasting of the educational programs of the Boston school department and his requirement that contracts for unwanted advice be given by the financially burdened Boston schools to the colleges and universities in the Boston area, because the judge and his appointed experts thought that such advice might help the process of desegregation through busing. We could give many other examples of action that might or might not fall under the general guidelines set by the Supreme Court in key decisions. But even in the United States, litigation has its limits. Many of these issues will never come to the Supreme Court—or an overburdened Court may refuse to deal with them when they do, and lower court decisions will stand as the law.

We might envisage this first factor—working out the logic of positions already taken—as a kind of indigestion, in which a boa constrictor, having swallowed a goat, must allow it to go through its entire length to be absorbed. So must these new expansions in "standing," due process, equal protection, and the like work their way through the entire system. They are far from having done so.

The Court is committed to an activist posture, with great impact on various areas of life, by the expansion of the reach of the legal principles on the basis of which it operates. Some assert, and in some angry dissents Justices themselves charge, that no legal logic guides the Court—that it is simply legislating its views on difficult problems. If the Court truly has cut loose from legal principles, one may envisage either that it may continue in an activist posture steadily, or that, with the appointment of new conservative judges, it will simply feel free to abandon the ground staked out by its predecessors and reverse precedents wholesale.

When government expands

But a second factor that sustains the permanent activism of the court is the enormous increase in the reach of government itself. When government expands, it could seem reasonable that the Court must extend its reach also. It must consider issues of equity and due process and equal protection in all the varied areas of education, health care, housing, and access to government services of all types. It must consider the varied impact of new subsidies, and controls and restrictions based on safety or environmental considerations. It is true that as government expands it sets up quasi-judicial bodies to adjudicate difficult decisions, but there is one major route of appeal in our system from these multifarious quasi-judicial bodies, and that is to the federal courts, and only one final appeal, to the Supreme Court. The expanded reach of government also means that new bases for decision-making by courts must be taken into account, that the "facts" on the basis of which lower courts rule become more and more complex. One murky realm opened up by this new complexity is the use of social science data—or perhaps some might say its misuse—as one side or another believes that the research findings of the moment support its position. To quote Archibald Cox again:

> Modern psychology has raised doubts concerning freedom of the will that raise skepticism of the very notion of crime. Sociologists have cast doubt upon the efficacy of punishment and deterrence in the face of the social, economic, and psychological causes of criminal conduct. When an issue is nicely balanced between the interests of the public and the claims of individual liberty, the substitution of such doubts for once-accepted verities may be enough to tip the scales against the prosecution.
>
> If this is true in criminal procedure, may not similar forces be partly responsible for the turmoil in other areas of constitutional law? . . . While the social scientists [are] changing our understanding of man . . . judges will inevitably be stimulated to reexamine the law's own presuppositions. One wonders, indeed, whether the gulf between the Supreme Court and the Congress is not partly a reflection of the closer kinship the justices have with the intellectual community.

However, as this passage will suggest to the 1975 reader, social science may provide successively new bases for new laws: Since the pattern of development in the social sciences is such that a set of findings is not established for long, social science may "require" new laws when the findings which support old laws are overturned. For instance, social scientists now think they may well be able to make a case for deterrence, and the fragmentary findings that supported some part of the decisions on school desegregation steadily have become weaker.

The expanded reach of government not only explains a more activist Court; in the minds of many analysts, it also justifies it. Perhaps it does. But one reason it does is that courts are dissatisfied with how legislatures and executives run their respective spheres, and while they do not egregiously reach out to express their dissatisfaction—courts, after all, must wait for cases to come to them—when the cases do come to them, they stretch their hands out very far indeed to make corrections. Consider issues raised in some recent cases: inadequate medical treatment for prisoners; welfare to applicants delayed beyond some reasonable time; public housing poorly maintained and in poor repair; mental patients not receiving treatment. The courts and their defenders say that if the legislature and executive are incapable of action in these and similar cases, then the courts must act.

Going to the root of problems

There is much justice here. Ward Elliott, who has sharply criticized the activist role of the Court in the reapportionment cases in *The Rise of Guardian Democracy,* agrees that in some states there was extreme malapportionment, about which something should have been done by someone. Increasingly, however, the courts have gone beyond the wrong presented to them to sweepingly reorganize a complex service of government so that the wrong can be dealt with —in the Court's mind, at least—at its root. Thus, a judge might decree, "Let this prisoner receive adequate medical care"—or, as he did instead, go to experts to provide a complete.program of medical services for prisons on the basis of what professionals asserted was necessary, a program which the state insisted it could not afford. A court could require that welfare recipients receive more rapid treatment, but what a federal judge in Massachusetts did was to suspend all federal government welfare reimbursements to the state until the state hired 255 more social workers.[2] If public hous-

---

[2] The Boston *Globe* reported (February 26, 1975) that the Legislature voted, under constraint to prevent the cutoff of federal welfare funds, to appropriate the funds to hire the social workers:

ing was demonstrated to be in poor repair, the judge could have said, "Fix it." But in this case, he enjoined the expenditure of other state funds for new housing, requiring that they be used for massive rehabilitation of old housing, appointed a master to determine the best way to repair and keep the house in repair, and suggested that the Boston Housing Authority didn't know how to create the kind of good community in which vandalism would not take place.[3] Similarly, judges are now determining with the aid of psychiatric experts what a proper system of psychiatric care in a mental hospital should be.

The justification in these and many other cases is that the legislature and executive won't act. This justification will not hold water. The legislature and executive have far more resources than the courts to determine how best to act. If they don't, it is because no one knows how to, or there is not enough money to cover everything, or because the people simply don't want it. These strike me as valid considerations in a democracy, but they are not considered valid considerations when issues of social policy come up as court cases for judgment. For example, no desegregation decision that I know of has been stayed by the fact that there is not enough money or that other school and educational services will suffer, perfectly valid considerations for legislatures, executives, and administrators —but the kind of consideration that no judge considers worthy of notice.

Having decided that the other two branches won't act, judges decide to act on their own, and increasingly are intrigued by the opportunity to go to the root of the problem. Unfortunately in many of these areas of social policy there is no clear knowledge of what the root of the problems is, though an expert can always be found who will oblige a judge with an appropriate program. A public health specialist will oblige the judge with a program for medical

United States District Judge Frank H. Freedman ordered the cutoff of Federal welfare funds—about $550 million a year. . . . But he said he would stay his order if the $1.2 million is approved this week. . . . He . . . threatened a new cutoff of funds if the state does not hire an additional 90 social workers before June 30. . . . The order for 90 more social workers, in addition to the 255, will force the Dukakis administration to file another deficiency budget. . . . Freedman also ordered that all new positions be carried over into the next fiscal year. In addition, he said that state must make permanent the 154 social worker job slots that are now covered under the federally-paid Comprehensive Employment Training Act (CETA).

[3] The Supreme Judicial Court of Massachusetts overruled the enjoining of state housing funds; what is interesting is that the lower court judge thought that he had the power to require state bond issues designed for one purpose to be used for another.

care that follows the standards of his professional association, standards that hardly any public body may be able to afford to meet or is interested to meet. And so with a psychiatrist, social work specialist, or school specialist. Clearly, if the judge has decided that the services in question are inadequate, or that they violate the constitution, or the laws, or the health code, or equal treatment, or whatever, he will find some expert who agrees with him.

Thus, the reach of government, already grossly expanded beyond its capacity to perform, is further expanded by the courts. Many elected officials now believe that government cannot deliver what has been promised in certain areas, either because of limited resources or knowledge; but it will be an interesting question whether the courts will now allow government to withdraw from these areas. Efforts to restrict welfare expenditure in those states where it has become a huge burden have been fought tooth and nail in the courts, and one may be sure that every other effort to withdraw from the provision of service will also be fought, by the professional groups providing the services, by the publicly-funded legal advocacy centers now established to protect the rights of various groups of citizens and recipients of government benefits, and by the beneficiaries themselves. Much will depend on the temper of the courts, and on the guidance the Supreme Court gives.

The new litigation

Finally, there is a third factor which suggests that a highly activist and intrusive judiciary is now a permanent part of the American Commonwealth: The courts will not be allowed to withdraw from the broadened positions they have seized, or have been forced to move into, because of the creation of new and powerful interests, chief among them the public advocacy law centers. It can hardly be an accident that the failure of the expected conservative cycle to succeed the activist cycle of 1954-68 occurred at the same time that many new centers were established for the promotion of social change through litigation. At the beginning of the 1960's there apparently existed only one such center—the NAACP Legal Defense Fund. Under the Economic Opportunity Act, many poverty law centers were created. Many other centers, receiving government or foundation aid, were established in almost every field of social policy—welfare, education, housing, health, environment—and for almost every group of potential clients—Mexican-Americans, Puerto Ricans, Indians, prisoners, mental health patients, etc. Law for the promotion of social change became enormously popular with law students, and many sought posts in the new centers.

Of course, this revolutionary change in the landscape of the practice of law itself reflected broader changes: a rising critical attitude toward government, a widespread belief among many sectors of the population in the unfairness and unjustness of government, the widespread legitimation among the youth and minorities of an adversary posture and denunciatory rhetoric—which all complemented nicely the standard practices of litigation. Law—for the purpose of the correction of presumed evils, for changing government practices, for overruling legislatures, executives, and administrators, for the purpose indeed of replacing democratic procedures with the authoritarian decisions of judges—became enormously popular. The number of law students rose rapidly, in response to new opportunities for litigation, and also serving as insurance of expanded litigation, owing to the increasing number of lawyers.

It is not easy to disentangle the complex web of elements that has created the powerful and permanent interests engaged in constitutional litigation to expand the scope and power of government; that these interests have been created, and have replaced those powerful business interests of the 1930's and 1940's that engaged in constitutional litigation to *restrict* the power of government, is not to be denied. In the social policy decisions of the courts, one sees many such interests at work: professional organizations, unions, clients, and recipients of benefits. Owing to changes in constitutional law, these interests now have greater access to the courts, concerning more grievances, than ever before. Public advocacy law firms may represent their interests, without the need for plaintiffs themselves to provide for their own legal fees. There is much in this situation which makes possible the separation of constitutional cases from any given client or interest. There are cases where the plaintiff seems to have disappeared—for example, the continuing effort to force the Southern states to eliminate the identifiably black character of their formerly segregated black colleges. But the cases lead a life of their own, with lawyers to argue them, with ghostly plaintiffs who never appear and, for all one knows, may not exist. Lawyers strive for objectives in the service of ideologies that cannot be realized through the legislative and executive functions of government but may be through the agency of authoritarian courts.

In all this there is much of the "guardian ethic" that Ward Elliott has characterized and criticized in his study of reapportionment. At first it was believed that a strict reapportionment of electoral districts on the basis of one man to one vote would shift some power to the cities or to the minorities who lived in them. Quite early it became clear it would not—at best it would help the conservative suburbs. In the event it turned out that it had almost no consequence at all. An ideological exercise was nevertheless carried out by those who thought they knew better, or who simply wanted to clean up the messiness of history and the past, unfortunately at the cost of stripping away the last shred of pretense that states had some degree of sovereignty, and by losing great stores of respect for the neutrality and objectivity of the courts. Other cases show the same insistence on uniformity that Elliott characterizes as part of the "guardian ethic": Thus, no one must pray or read the Bible in any school, even when there is no one to object.

But if the "guardian ethic" often became dissociated from any concrete interest, one must recognize that these concrete interests do exist. One must also recognize that they were encouraged to come into being and to grow by the explosion of publicly supported litigation for the establishment of new rights. People unaware of grievances—women, or prisoners, or welfare recipients, or Mexican-Americans—were educated to feel them by those who seemed to divine what those grievances should be, and even if they were insufficiently concerned to act on them, lawyers were readily available, paid for by government or foundations, to take up their interests. Moral fervor and outrage, properly aroused by great inequalities in American life—in particular, the legal and extra-legal subordination of Negroes—were transferred in the course of the 1960's to practices that most citizens did not think iniquitous or outrageous or improper at all.

Of course, people will disagree where iniquity ends and when moral outrage is unjustified. Some will be satisfied when the right to vote is guaranteed, others only when literacy laws are suspended, electoral literature is translated into any language a voter may know, and voters form a perfect statistical cross section, by race and ethnicity and age, of the population. Some will be satisfied when discrimination is outlawed, others only when quotas by ethnicity and sex are set for every job. Some will be satisfied when prisoners and mental patients are not abused, others only when the procedures that they think will lead to rehabilitation or cure are imposed by the courts. The law, generally made by judges but with the assistance also of legislatures and administrators, has moved insensibly from the first of each of these alternatives, which is as far as anyone wanted to go 10 years ago, to the second.

The trust of the people

We may debate whether we have a better society or commonwealth or a worse one as a result. I believe we have a considerably
worse one, because a free people feels itself increasingly under the
arbitrary rule of unreachable authorities, and that cannot be good
for the future of the state. Even the guardians of the "guardian
ethic"—the better educated, the establishment, the opinion-makers—
are now doubtful of many of the rulings they urged when, unable
to institute them through the elected representatives of the people,
they made law through recourse to the courts. But in the meantime
the great fund of respect and trust by the people for governmental
institutions has been drawn down; the courts, trying to create a better society, have increasingly lost the respect and trust of the people—which in the end is what sustained and must sustain the remarkable institution of a supreme judiciary in American life. Even
in 1965, McCloskey could write, "'Judicial realism' has eroded the
traditional mystique that often lent authority in other days. . . ."
That trust is considerably further eroded 10 years later. "To con-
strue the law," Bryce wrote of the Supreme Court, "that is, to elu-
cidate the will of the people as supreme lawgiver, is the beginning
and end of their duty." Bryce then adds in a footnote:

> 'Suppose, however,' someone may say 'that the court should go beyond
> its duty and import its own views of what ought to be the law into its
> decision as to what is the law. This would be an exercise of judicial
> will.' Doubtless, it would, but it would be a breach of duty, would
> expose the court to the distrust of the people, and might, if repeated
> or persisted in a serious matter, provoke resistance to the law as laid
> down by the court.

We have seen efforts at resistance: All have failed, and there is
little enough now. Some may see this as the triumph of law instituting justice; I suspect it is rather the apathy of cynical and baffled
people, incapable of seeing what actions may release them from the
toils of the intrusive courts.

A final quotation from Bryce:

> In America the Constitution is at all times very hard to change: much
> more then must political issues turn on its interpretation. And if this
> be so, must not the interpreting court be led to assume a control over
> the executive and legislative branches of government, since it has the
> power of declaring their acts illegal?
>
> There is ground for these criticisms. The evil they point to has oc-
> curred and may recur. But it occurs very rarely, and may be averted
> by the same prudence which the courts have hitherto generally shown.

In 1954 the Court abandoned prudence and for 15 years firmly and unanimously insisted that the segregation and degradation of the Negro must end. It succeeded, and eventually the legislative and executive branches came to its side and the heritage of unequal laws and unequal treatment was eliminated. That was indeed a heroic period in the history of the court. But even heroes may overreach themselves. It is now time for the Court to act with the prudence that must in a free society be the more regular accompaniment of its actions.

6

On
corporate capitalism
in
America

IRVING KRISTOL

T HE United States is the capitalist nation *par excellence*. That is to say, it is not merely the case that capitalism has flourished here more vigorously than, for instance, in the nations of Western Europe. The point is, rather, that the Founding Fathers *intended* this nation to be capitalist and regarded it as the *only* set of economic arrangements consistent with the liberal democracy they had established. They did not use the term "capitalism," of course; but, then, neither did Adam Smith, whose *Wealth of Nations* was also published in 1776, and who spoke of "the system of natural liberty." That invidious word, "capitalism," was invented by European socialists about a half-century later—just as our other common expression, "free enterprise," was invented still later by anti-socialists who saw no good reason for permitting their enemies to appropriate the vocabulary of public discourse. But words aside, it is a fact that capitalism in this country has a historical legitimacy that it does not possess elsewhere. In other lands, the nation and its fundamental institutions antedate the capitalist era; in the United States, where liberal democracy is not merely a form of government but also a "way of life," capitalism and democracy have been organically linked.

This fact, quite simply accepted until the 1930's—accepted by both radical critics and staunch defenders of the American regime—has been obscured in recent decades by the efforts of liberal scholars to create a respectable pedigree for the emerging "welfare state." The impetus behind this scholarship was justified, to a degree. It is true that the Founding Fathers were not dogmatic *laisser-fairists,* in a later neo-Darwinian or "libertarian" sense of the term. They were intensely suspicious of governmental power, but they never could have subscribed to the doctrine of "our enemy, the State." They believed there was room for some governmental intervention in economic affairs; and—what is less frequently remarked—they believed most firmly in the propriety of governmental intervention and regulation in the areas of public taste and public morality. But, when one has said this, one must add emphatically that there really is little doubt that the Founders were convinced that economics was the sphere of human activity where government intervention was, as a general rule, least likely to be productive, and that "the system of natural liberty" in economic affairs was the complement to our system of constitutional liberty in political and civil affairs. They surely would have agreed with Hayek that the paternalistic government favored by modern liberalism led down the "road to serfdom."

But one must also concede that both the Founding Fathers and Adam Smith would have been perplexed by the kind of capitalism we have in 1976. They could not have interpreted the domination of economic activity by large corporate bureaucracies as representing, in any sense, the working of a "system of natural liberty." Entrepreneurial capitalism, as they understood it, was mainly an individual—or at most, a family—affair. Such large organizations as might exist—joint stock companies, for example—were limited in purpose (e.g., building a canal or a railroad) and usually in duration as well. The large, publicly-owned corporation of today which strives for immortality, which is committed to no line of business but rather (like an investment banker) seeks the best return on investment, which is governed by an anonymous oligarchy—such an institution would have troubled and puzzled them, just as it troubles and puzzles us. And they would have asked themselves the same questions we have been asking ourselves for almost a century now: Who "owns" this new leviathan? Who governs it—and by what right, and according to what principles?

The unpopular revolution

To understand the history of corporate capitalism in America, it is important to realize in what sense it may be fairly described as an "accidental institution." Not in the economic sense, of course. In the latter part of the last century, in all industrialized nations, the large corporation was born out of both economic necessity and economic opportunity: the necessity of large pools of capital and of a variety of technical expertise to exploit the emerging technologies, and the opportunity for economies of scale in production, marketing, and service in a rapidly-urbanizing society. It all happened so quickly that the term "corporate revolution" is not inappropriate. In 1870, the United States was a land of small family-owned business. By 1905, the large, publicly-owned corporation dominated the economic scene.

But the corporate revolution was always, during that period, an unpopular revolution. It was seen by most Americans as an accident of economic circumstance—something that happened to them rather than something they had created. They had not foreseen it; they did not understand it—in no way did it seem to "fit" into the accepted ideology of the American democracy. No other institution in American history—not even slavery—has ever been so consistently unpopular as has the large corporation with the American public. It was controversial from the outset, and it has remained controversial to this day.

This is something the current crop of corporate executives find very difficult to appreciate. Most of them reached maturity during the post-war period, 1945-1960. As it happens, this was—with the possible exception of the 1920's—just about the only period when public opinion was, on the whole, well-disposed to the large corporation. After 15 years of depression and war, the American people wanted houses, consumer goods, and relative security of employment—all the things that the modern corporation is so good at supplying. The typical corporate executive of today, in his 50's or 60's, was led to think that such popular acceptance was "normal," and is therefore inclined to believe that there are novel and specific forces behind the upsurge of anti-corporate sentiment in the past decade. As a matter of fact, he is partly right: There *is* something significantly new about the hostility to the large corporation in our day. But there is also something very old, something coeval with the very existence of the large corporation itself. And it is the interaction of the old hostility with the new which has put the modern corporation in the critical condition that we find it in today.

The old hostility is based on what we familiarly call "populism." This is a sentiment basic to any democracy—indispensable to its establishment but also, ironically, inimical to its survival. Populism is the constant fear and suspicion that power and/or authority, whether in government or out, is being used to frustrate "the will of the people." It is a spirit that intimidates authority and provides the popular energy to curb and resist it. The very possibility of a democratic society—as distinct from the forms of representative government, which are its political expression—is derived from, and is constantly renewed by, the populist temper. The Constitution endows the United States with a republican form of government, in which the free and explicit consent of the people must ultimately ratify the actions of those in authority. But the populist spirit, which both antedated and survived the Constitutional Convention, made the United States a democratic nation as well as a republican one—committed to "the democratic way of life" as well as to the proprieties of constitutional government. It is precisely the strength of that commitment which has always made the American democracy somehow different from the democracies of Western Europe—a difference which every European observer has been quick to remark.

But populism is, at the same time, an eternal problem for the American democratic republic. It incarnates an antinomian impulse, a Jacobin contempt for the "mere" forms of law and order and civility. It also engenders an impulse toward a rather infantile political utopianism, on the premise that nothing is too good for "the people." Above all, it is a temper and state of mind which too easily degenerates into political paranoia, with "enemies of the people" being constantly discovered and exorcised and convulsively purged. Populist paranoia is always busy subverting the very institutions and authorities that the democratic republic laboriously creates for the purpose of orderly self-government.

In the case of the large corporation, we see a healthy populism and a feverish paranoia simultaneously being provoked by its sudden and dramatic appearance. The paranoia takes the form of an instinctive readiness to believe anything reprehensible, no matter how incredible, about the machinations of "big business." That species of journalism and scholarship which we call "muckraking" has made this kind of populist paranoia a permanent feature of American intellectual and public life. Though the businessman *per se* has never been a fictional hero of bourgeois society (as Stendhal observed, a merchant may be honorable but there is nothing heroic about him), it is only after the rise of "big business" that the businessman becomes the natural and predestined villain of the novel, the drama, the cinema, and, more recently, television. By now most Americans are utterly convinced that all "big business" owes its existence to the original depredations of "robber barons"—a myth which never really was plausible, which more recent scholarship by economic historians has thoroughly discredited, but which probably forever will have a secure hold on the American political imagination. Similarly, most Americans are now quick to believe that "big business" conspires secretly but most effectively to manipulate the economic and political system—an enterprise which, in prosaic fact, corporate executives are too distracted and too unimaginative even to contemplate.

Along with this kind of paranoia, however, populist hostility toward the large corporation derives from an authentic bewilderment and concern about the place of this new institution in American life. In its concentration of assets and power—power to make economic decisions affecting the lives of tens of thousands of citizens—it seemed to create a dangerous disharmony between the economic system and the political. In the America of the 1890's, even government did not have, and did not claim, such power (except in wartime). *No one* was supposed to have such power—it was, indeed, a radical diffusion of power that was thought to be an essential characteristic of democratic capitalism. The rebellion of Jacksonian democracy against the Bank of the United States had been directed precisely against such an "improper" concentration of power. A comparable rebellion now took place against "big business."

"Big business" or capitalism?

It was not, however, a rebellion against capitalism as such. On the contrary, popular hostility to the large corporation reflected the fear that this new institution was subverting capitalism as Americans then understood (and, for the most part, still understand) it. This understanding was phrased in individualistic terms. The entrepreneur was conceived of as a real person, not as a legal fiction. The "firm" was identified with such a real person (or a family of real persons) who took personal risks, reaped personal rewards, and assumed personal responsibility for his actions. One of the consequences of the victorious revolt against the Bank of the United States had been to make the chartering of corporations—legal "persons" with limited liability—under state law a routine and easy thing, the assumption being that this would lead to a proliferation of small corporations, still easily identifiable with the flesh-and-blood entrepreneurs who founded them. The rise of "big business" frustrated such expectations.

Moreover, the large corporation not only seemed to be but actually was a significant deviation from traditional capitalism. One of the features of the large corporation—though more a consequence of its existence than its cause—was its need for, and its ability to create, "orderly markets." What businessmen disparagingly call "cutthroat competition," with its wild swings in price, its large fluctuations in employment, its unpredictable effects upon profits —all this violates the very *raison d'être* of a large corporation, with its need for relative stability so that its long-range investment decisions can be rationally calculated. The modern corporation always looks to the largest and most powerful firm in the industry to establish "market leadership" in price, after which competition will concentrate on quality, service, and the introduction of new products. One should not exaggerate the degree to which the large corporation is successful in these efforts. John Kenneth Galbraith's notion that the large corporation simply manipulates its market through the power of advertising and fixes the price level with sovereign authority is a wild exaggeration. This is what all corporations *try* to do; it is what a few corporations, in some industries, sometimes succeed in doing. Still, there is little doubt that the idea of a "free market," in the era of large corporations, is not quite the original capitalist idea.

The populist response to the transformation of capitalism by the large corporation was, and is: "Break it up!" Anti-trust and anti-monopoly legislation was the consequence. Such legislation is still enacted and re-enacted, and anti-trust prosecutions still make headlines. But the effort is by now routine, random, and largely pointless. There may be a few lawyers left in the Justice Department or the Federal Trade Commission who sincerely believe that such laws, if stringently enforced, could restore capitalism to something like its pristine individualist form. But it is much more probable that the lawyers who staff such government agencies launch these intermittent crusades against "monopoly" and "oligopoly"—terms that are distressingly vague and inadequate when applied to the real world—because they prefer such activity to mere idleness, and because they anticipate that a successful prosecution will enhance their professional reputations. No one expects them to be effectual, whether the government wins or loses. Just how much difference, after all, would it make if AT&T were forced to spin off its Western Electric manufacturing subsidiary, or if IBM were divided into three different computer companies? All that would be accomplished is a slight increase in the number of large corporations, with very little consequence for the shape of the economy or the society as a whole.

True, one could imagine—in the abstract—a much more radical effort to break up "big business." But there are good reasons why, though many talk solemnly about this possibility, no one does anything about it. The costs would simply be too high. The economic costs, most obviously: an adverse effect on productivity, on capital investment, on our balance of payments, etc. But the social and political costs would be even more intolerable. Our major trade unions, having after many years succeeded in establishing collective bargaining on a national level with the large corporation, are not about to sit back and watch their power disintegrate for the sake of an ideal such as "decentralization." And the nation's pension funds are not about to permit the assets of the corporations in which they have invested to be dispersed, and the security of their pension payments correspondingly threatened.

One suspects that even popular opinion, receptive in principle to the diminution of "big business," would in actuality find the process too painful to tolerate. For the plain fact is that, despite much academic agitation about the horrors of being an "organization man," the majority of those who now work for a living, of whatever class, have learned to prefer the security, the finely-calibrated opportunities for advancement, the fringe benefits, and the paternalism of a large corporation to the presumed advantages of employment in smaller firms. It is not only corporate executives who are fearful of "cutthroat competition"; most of us, however firmly we declare our faith in capitalism and "free enterprise," are sufficiently conservative in our instincts to wish to avoid all such capitalist rigors. Even radical professors, who in their books find large bureaucratic corporations "dehumanizing," are notoriously reluctant to give up tenured appointments in large bureaucratic universities for riskier opportunities elsewhere.

So the populist temper and the large corporation coexist uneasily in America today, in what can only be called a marriage of convenience. There is little affection, much nagging and backbiting and whining on all sides, but it endures—"for the sake of the children," as it were. Not too long ago, there was reason to hope that, out of the habit of coexistence, there would emerge something like a philosophy of coexistence: a mutual adaptation of the democratic-individualist-capitalist ideal and the bureaucratic-corporate reality, sanctioned by a new revised version of the theory of democracy and capitalism—a new political and social philosophy, in short, which extended the reach of traditional views without repudiating them. But that possibility, if it was ever more than a fancy, has been effectively cancelled by the rise, over the past decade, of an anti-capitalist ethos which has completely transformed the very definition of the problem.

The anti-liberal Left

This ethos, in its American form, is not *explicitly* anti-capitalistic, and this obscures our perception and understanding of it. It has its roots in the tradition of "Progressive-reform," a tradition which slightly antedated the corporate revolution but which was immensely stimulated by it. In contrast to populism, this was (and is) an upper-middle class tradition—an "elitist" tradition, as one would now say. Though it absorbed a great many socialist and neo-socialist and quasi-socialist ideas, it was too American—too habituated to the rhetoric of individualism, and even in some measure to its reality —to embrace easily a synoptic, collectivist vision of the future as enunciated in socialist dogmas. It was willing to contemplate "public ownership" (i.e., ownership by the political authorities) of *some* of the "means of production," but on the whole it preferred to think in terms of *regulating* the large corporation rather than nationalizing it or breaking it up. It is fair to call it an indigenous and peculiarly American counterpart to European socialism—addressing itself to the same problems defined in much the same way, motivated by the same ideological impulse, but assuming an adversary posture toward "big business" specifically rather than toward capitalism in general.

At least, that is what "Progressive-reform" used to be. In the past decade, however, it has experienced a transmutation of ideological substance while preserving most of the traditional rhetorical wrappings. That is because it embraced, during these years, a couple of other political traditions, European in origin, so that what we still call "liberalism" in the United States is now something quite different from the liberalism of the older "Progressive-reform" impulse. It is so different, indeed, as to have created a cleavage between those who think of themselves as "old liberals"—and are now often redesignated as "neo-conservatives"—and the new liberals who are in truth men and women of "the Left," in the European sense of that term. This is an important point, worthy of some elaboration and clarification—especially since the new liberalism is not usually very candid about the matter.[1]

The Left in Europe, whether "totalitarian" or "democratic," has consistently been anti-liberal. That is to say, it vigorously repudiates the intellectual traditions of liberalism—as expressed, say, by Locke, Montesquieu, Adam Smith, and Tocqueville—and with equal vigor rejects the key institution of liberalism: the (relatively) free market (which necessarily implies limited government). The Left emerges out of a rebellion against the "anarchy" and "vulgarity" of a civilization that is shaped by individuals engaged in market transactions. The "anarchy" to which it refers is the absence of any transcending goal or purpose which society is constrained to pursue —and which socialists, with their superior understanding of History, feel obligated to prescribe. Such a prescription, when fulfilled, will supposedly reestablish a humane "order." The "vulgarity" to which it refers is the fact that a free market responds—or tries to respond—to the appetites and preferences of common men and women, whose use of their purchasing power determines the shape of the civilization. Since common men and women are likely to have "common" preferences, tastes, and aspirations, the society they create—the "consumption society," as it is now called—will be regarded by some critics as short-sightedly "materialistic." People will seek to acquire what they want (e.g., automobiles), not why they "need" (e.g., mass transit). Socialists are persuaded that they have a superior understanding of people's true needs, and that the people will be more truly happy in a society where socialists have the authority to define those needs, officially and unequivocally.

Obviously, socialism is an "elitist" movement, and in its beginnings—with Saint-Simon and Auguste Comte—was frankly conceived of as such. Its appeal has always been to "intellectuals" (who feel dispossessed by and alienated from a society in which they are merely one species of common man) and members of the upper-middle class who, having reaped the benefits of capitalism, are now in a position to see its costs. (It must be said that these costs are not imaginary: Socialism would not have such widespread appeal if its critique of liberal capitalism were entirely without

[1] It must be said, however, that even when it is candid, no one seems to pay attention. John Kenneth Galbraith has recently publicly defined himself as a "socialist," and asserts that he has been one—whether wittingly or unwittingly, it is not clear—for many years. But the media still consistently identify him as a "liberal," and he is so generally regarded. Whether this is mere habit or instinctive protection coloration—for the media are a crucial wing of the "new liberalism"—it is hard to say.

substance.) But all social movements in the modern world must define themselves as "democratic," since democratic legitimacy is the only kind of legitimacy we recognize. So "totalitarian" social-ism insists that it is a "people's democracy," in which the "will of the people" is mystically incarnated in the ruling party. "Demo-cratic socialism," on the other hand, would like to think that it can "socialize" the economic sector while leaving the rest of society "liberal." As Robert Nozick puts it, democratic socialists want to proscribe only *"capitalist* transactions between consenting adults."

The trouble with the latter approach is that democratic social-ists, when elected to office, discover that to collectivize economic life you have to coerce all sorts of other institutions (e.g. the trade unions, the media, the educational system) and limit individual free-dom in all sorts of ways (e.g., freedom to travel, freedom to "drop out" from the world of work, freedom to choose the kind of edu-cation one prefers) if a "planned society" is to function efficiently. When "democratic socialist" governments show reluctance to take such actions, they are pushed into doing so by the "left wings" of their "movements," who feel betrayed by the distance that still exists between the reality they experience and the socialist ideal which enchants them. Something like this is now happening in all the European social-democratic parties and in a country like India.

The "new class"

The United States never really had any such movement of the Left, at least not to any significant degree. It was regarded as an "un-American" thing, as indeed it was. True, the movement of "Progressive-reform" was "elitist" both in its social composition and its social aims: It, too, was distressed by the "anarchy" and "vul-garity" of capitalist civilization. But in the main it accepted as a fact the proposition that capitalism and liberalism were organical-ly connected, and it proposed to itself the goal of "mitigating the evils of capitalism," rather than abolishing liberal capitalism and replacing it with "a new social order" in which a whole new set of human relationships would be established. It was an authentic *reformist* movement. It wanted to regulate the large corporations so that this concentration of private power could not develop into an oligarchical threat to democratic-liberal-capitalism. It was ready to interfere with the free market so that the instabilities generated by capitalism—above all, instability of employment—would be less costly in human terms. It was even willing to tamper occasionally with the consumer's freedom of choice where there was a clear consensus that the micro-decisions of the marketplace added up to macro-consequences that were felt to be unacceptable. And it hoped to correct the "vulgarity" of capitalist civilization by educating the people so that their "preference schedules" (as economists would say) would be, in traditional terms, more elevated, more appreciative of "the finer things in life."

Ironically, it was the extraordinary increase in mass higher education after World War II that, perhaps more than anything else, infused the traditional movement for "Progressive-reform" with various impulses derived from the European Left. The earlier movement had been "elitist" in fact as well as in intention—i.e., it was sufficiently small so that, even while influential, it could hardly contemplate the possibility of actually exercising "power." Mass higher education has converted this movement into something like a mass movement proper, capable of driving a President from office (1968) and nominating its own candidate (1972). The intentions remain "elitist," of course; but the movement, under the banner of "the New Politics," now encompasses some millions of people. These are the people whom liberal capitalism had sent to college in order to help manage its affluent, highly technological, mildly paternalistic, "post-industrial" society.

This "new class" consists of scientists, lawyers, city planners, social workers, educators, criminologists, sociologists, public health doctors, etc.—a substantial number of whom find their careers in the expanding public sector rather than the private. The public sector, indeed, is where they prefer to be. They are, as one says, "idealistic"—i.e., far less interested in individual financial rewards than in the corporate power of their class. Though they continue to speak the language of "Progressive-reform," in actuality they are acting upon a hidden agenda: to propel the nation from that modified version of capitalism we call "the welfare state" toward an economic system so stringently regulated in detail as to fulfill many of the traditional anti-capitalist aspirations of the Left.

The exact nature of what has been happening is obscured by the fact that this "new class" is not merely liberal but truly "libertarian" in its approach to all areas of life—except economics. It celebrates individual liberty of speech and expression and action to an unprecedented degree, so that at times it seems almost anarchistic in its conception of the good life. But this joyful individualism always stops short of the border where economics—i.e., capitalism —begins. The "new class" is surely sincere in such a contradictory commitment to a maximum of individual freedom in a society where economic life becomes less free with every passing year. But it is instructive to note that these same people, who are irked and inflamed by the slightest non-economic restriction in the United States, are quite admiring of Maoist China and not in the least appalled by the total collectivization of life—and the total destruction of liberty—there. They see this regime as "progressive," not "reactionary." And, in this perception, they unwittingly tell us much about their deepest fantasies and the respective quality of their political imagination.

Meanwhile, the transformation of American capitalism proceeds apace. Under the guise of coping with nasty "externalities"—air pollution, water pollution, noise pollution, traffic pollution, health pollution, or what have you—more and more of the basic economic decisions are being removed from the marketplace and transferred to the "public"—i.e., political—sector, where the "new class," by virtue of its expertise and skills, is so well represented. This movement is naturally applauded by the media, which are also for the most part populated by members of this "new class" who believe —as the Left has always believed—it is government's responsibility to cure all the ills of the human condition, and who ridicule those politicians who deny the possibility (and therefore the propriety) of government doing any such ambitious thing. And, inevitably, more explicitly socialist and neo-socialist themes are beginning boldly to emerge from the protective shell of Reformist-liberal rhetoric. The need for some kind of "national economic plan" is now being discussed seriously in Congressional circles; the desirability of "public"—i.e., political—appointees to the boards of directors of the largest corporations is becoming more apparent to more politicians and journalists with every passing day; the utter "reasonableness," in principle, of price and wage controls is no longer even a matter for argument, but is subject only to circumstantial and prudential considerations. Gradually, the traditions of the Left are being absorbed into the agenda of "Progressive-reform," and the structure of American society is being radically, if discreetly, altered.

"The enemy of being is having"

One of the reasons this process is so powerful, and meets only relatively feeble resistance, is that it has a continuing source of energy within the capitalist system itself. That source is not the "inequalities" or "injustices" of capitalism, as various ideologies of the Left insist. These may represent foci around which dissent is occasionally and skillfully mobilized. But the most striking fact about anti-capitalism is the degree to which it is *not* a spontaneous working-class phenomenon. Capitalism, like all economic and social systems, breeds its own peculiar discontents—but the discontents of the working class are, in and of themselves, not one of its major problems. Yes, there is class conflict in capitalism; there is always class conflict, and the very notion of a possible society without class conflict is one of socialism's most bizarre fantasies. (Indeed, it is this fantasy that is socialism's original contribution to modern political theory; the importance of class conflict itself was expounded by Aristotle and was never doubted by anyone who ever bothered to look at the real world.) But there is no case, in any country that can reasonably be called "capitalist," of such class conflict leading to a proletarian revolution. Capitalism, precisely because its aim is the satisfaction of "common" appetites and aspirations, can adequately cope with its own class conflicts, through economic growth primarily and some version of the welfare state secondarily. It can do so, however, only if it is permitted to—a permission which the anti-capitalist spirit is loathe to concede. This spirit *wants* to see capitalism falter and fail.

The essence of this spirit is to be found, not in *The Communist Manifesto,* but rather in the young Marx who wrote: *"The enemy of being is having."* This sums up neatly the animus which intellectuals from the beginning, and "the new class" in our own day, have felt toward the system of liberal capitalism. This system is in truth "an acquisitive society," by traditional standards. Not that men and women under capitalism are "greedier" than under feudalism or socialism or whatever. Almost all people, almost all of the time, want more than they have. But capitalism is unique among social and economic systems in being organized for the overriding purpose of giving them more than they have. And here is where it runs into trouble: Those who benefit most from capitalism—and their children, especially—experience a withering away of the acquisitive impulse. Or, to put it more accurately: They cease to think of acquiring money and begin to think of acquiring power so as to improve the "quality of life," and to give *being* priority over *having.* That is the meaning of the well-known statement by a student radical of the 1960's: "You don't know what hell is like unless you were raised in Scarsdale." Since it is the ambition of capitalism to enable everyone to live in Scarsdale or its equivalent, this challenge is far more fundamental than the orthodox Marxist one, which says—against all the evidence—that capitalism will fail because it *cannot* get everyone to live in Scarsdale.

Against this new kind of attack, any version of capitalism would be vulnerable. But the version of corporate capitalism under which we live is not merely vulnerable; it is practically defenseless. It is not really hard to make a decent case, on a pragmatic level, for liberal capitalism today—especially since the anti-capitalist societies the 20th century has given birth to are, even by their own standards, monstrous abortions and "betrayals" of their originating deals. And corporate capitalism does have the great merit of being willing to provide a milieu of comfortable liberty—in universities, for example—for those who prefer *being* to *having*. But the trouble with the large corporation today is that it does not possess a clear theoretical—i.e., ideological—legitimacy within the framework of liberal capitalism itself. Consequently the gradual usurpation of managerial authority by the "new class"—mainly through the transfer of this authority to the new breed of regulatory officials (who are the very prototype of the class)—is almost irresistible.

Bureaucratic enterprise

So long as business was an activity carried on by real individuals who "owned" the property they managed, the politicians, the courts, and public opinion were all reasonably respectful of the capitalist proprieties. Not only was the businessman no threat to liberal democracy; he was, on the contrary, the very epitome of the bourgeois liberal-democratic ethos—the man who succeeded by diligence, enterprise, sobriety and all those other virtues that Benjamin Franklin catalogued for us, and which we loosely call "the Protestant Ethic." [2]

On the whole, even today, politicians and public opinion are inclined to look with some benevolence on "small business," and no one seems to be interested in leading a crusade against it. But the professionally-managed large corporation is another matter entirely. The top executives of these enormous bureaucratic institutions are utterly sincere when they claim fealty to "free enterprise," and they even have a point: Managing a business corporation, as distinct from a government agency, does require a substantial degree of entrepreneurial risk-taking and entrepreneurial skill. But it is also

---

[2] I say "loosely call" because, as a Jew, I was raised to think that this was an ancient "Hebrew ethic," and some Chinese scholars I have spoken to feel that it could appropriately be called "The Confucian ethic."

the case that they are as much functionaries as entrepreneurs, and rather anonymous functionaries at that. Not only don't we know who the chairman of General Motors is; we know so little about the kind of person who holds such a position that we haven't the faintest idea as to whether or not we want our children to grow up like him. Horatio Alger, writing in the era of pre-corporate capitalism, had no such problems. And there is something decidedly odd about a society in which a whole class of Very Important People is not automatically held up as one possible model of emulation for the young—and cannot be so held up because they are, as persons, close to invisible.

Nor is it at all clear whose interests these entrepreneur-functionaries are serving. In theory, they are elected representatives of the stockholder-"owners." But stockholder elections are almost invariably routine affirmations of management's will, because management will have previously secured the support of the largest stockholders; and for a long while now stockholders have essentially regarded themselves, and are regarded by management, as little more than possessors of a variable-income security. A stock certificate has become a lien against the company's earnings and assets—a subordinated lien, in both law and fact—rather than a charter of "citizenship" within a corporate community. And though management will talk piously, when it serves its purposes, about its obligations to the stockholders, the truth is that it prefers to have as little to do with them as possible, since their immediate demands are only too likely to conflict with management's long-term corporate plans.

It is interesting to note that when such an organization of business executives as the Committee on Economic Development drew up a kind of official declaration of the responsibilities of management a few years ago, it conceived of the professional manager as "a trustee balancing the interests of many diverse participants and constituents in the enterprise," and then enumerated these participants and constituents: employees, customers, suppliers, stockholders, government—i.e., practically everyone. Such a declaration serves only to ratify an accomplished fact: The large corporation has ceased being a species of private property, and is now a "quasi-public" institution. But if it is a "quasi-public" institution, some novel questions may be properly addressed to it: By what right does the self-perpetuating oligarchy that constitutes "management" exercise its powers? On what principles does it do so? To these essentially political questions management can only respond with the weak economic answer that its legitimacy derives from the superior efficiency with which it responds to signals from the free market. But such an argument from efficiency is not compelling when offered by a "quasi-public" institution. In a democratic republic such as ours, public and quasi-public institutions are not supposed simply to be efficient at responding to people's transient desires, are not supposed to be simply *pandering* institutions—but are rather supposed to help shape the people's wishes, and ultimately, the people's character, according to some version—accepted by the people itself—of the "public good" and "public interest." This latter task the "new class" feels itself supremely qualified to perform, leaving corporate management in the position of arguing that it is improper for this "quasi-public" institution to do more than give the people what they want—a debased version of the democratic idea which has some temporary demagogic appeal but no permanent force.

The corporation and liberal democracy

Whether for good or evil—and one can leave this for future historians to debate—the large corporation has gone "quasi-public," i.e., it now straddles, uncomfortably and uncertainly, both the private and public sectors of our "mixed economy." In a sense one can say that the modern large corporation stands to the bourgeois-individualist capitalism of yesteryear as the "imperial" American polity stands to the isolated republic from which it emerged: Such a development may or may not represent "progress," but there is no turning back.

The danger which this situation poses for the American democracy is not the tantalizing ambiguities inherent in such a condition —it is the genius of a pluralist democracy to convert such ambiguities into possible sources of institutional creativity and to avoid "solving" them, as a Jacobin democracy would, with one swift stroke of the sword. The danger is rather that the large corporation will be thoroughly integrated into the public sector, and lose its private character altogether. The transformation of American capitalism that *this* would represent—a radical departure from the quasi-bourgeois "mixed economy" to a system that could be fairly described as kind of "state capitalism"—does constitute a huge potential threat to the individual liberties Americans have traditionally enjoyed.

One need not, therefore, be an admirer of the large corporation to be concerned about its future. One might even regard its "bureaucratic-acquisitive" ethos, in contrast to the older "bourgeois-moralistic" ethos, as a sign of cultural decadence—and still be concerned about its future. In our pluralistic society we frequently find ourselves defending specific concentrations of power, about which we might otherwise have the most mixed feelings, on the grounds that they contribute to a general diffusion of power, a diffusion which creates the "space" in which individual liberty can survive and prosper. This is certainly our experience vis-à-vis certain religious organizations—e.g., the Catholic Church, the Mormons—whose structure and values are, in some respects at least, at variance with our common democratic beliefs, and yet whose existence serves to preserve our democracy as a free and liberal society. The general principle of checks and balances, and of decentralized authority too, is as crucial to the social and economic structures of a liberal democracy as to its political structure.

Nevertheless, it seems clear that the large corporation is not going to be able to withstand those forces pulling and pushing it into the political sector unless it confronts the reality of its predicament and adapts itself to this reality in a self-preserving way. There is bound to be disagreement as to the forms such adaptation should take, some favoring institutional changes that emphasize and clarify the corporation's "public" nature, others insisting that its "private" character must be stressed anew. Probably a mixture of both strategies would be most effective. If large corporations are to avoid having government-appointed directors on their boards, they will have to take the initiative and try to preempt that possibility by themselves appointing distinguished "outside" directors—directors from outside the business community. At the same time, if corporations are going to be able to resist the total usurpation of their decision-making powers by government, they must create a constituency—of their stockholders, above all—which will candidly intervene in the "political game" of interest-group politics, an intervention fully in accord with the principles of our democratic system.

In both cases, the first step will have to be to persuade corporate management that some such change is necessary. This will be difficult: Corporate managers are (and enjoy being) essentially economic-decision-making animals, and they are profoundly resentful of the "distractions" which "outside interference" of any kind will impose on them. After all, most chief executives have a tenure of about six years, and they all wish to establish the best possible track record, in terms of "bottom line" results, during that period. Very few are in a position to, and even fewer have an inclination to, take a long and larger view of the corporation and its institutional problems.

At the same time, the crusade against the corporations continues, with the "new class" successfully appealing to populist anxieties, seeking to run the country in the "right" way, and to reshape our civilization along lines superior to those established by the market-place. Like all crusades, it engenders an enthusiastic paranoia about the nature of the Enemy and the deviousness of His operations. Thus, the *New Yorker,* which has become the liberal-chic organ of the "new class," has discovered the maleficent potential of the mul-ti-national corporation at exactly the time when the multi-national corporation is in full retreat before the forces of nationalism every-where. And the fact that American corporations sometimes have to bribe foreign politicians—for whom bribery is a way of life—is in-flated into a rabid indictment of the personal morals of corporate executives. (That such bribery is also inherent in government-aid programs to the underdeveloped countries is, on the other hand, *never* taken to reflect on those—e.g., the World Bank—who institute and run such programs, and is thought to be irrelevant to the de-sirability or success of the program themselves.) So far, this crusade has been immensely effective. It will continue to be effective until the corporation has decided what kind of institution it is in today's world, and what kinds of reforms are a necessary precondition to a vigorous defense—not of its every action but of its very survival as a quasi-public institution as distinct from a completely politi-cized institution.

It is no exaggeration to say that the future of liberal democracy in America is intimately involved with these prospects for survival —the survival of an institution which liberal democracy never en-visaged, whose birth and existence have been exceedingly trouble-some to it, and whose legitimacy it has always found dubious. One can, if one wishes, call this a paradox. Or one can simply say that everything, including liberal democracy, is what it naturally be-comes—is what it naturally evolves into—and our problem derives from a reluctance to revise yesteryear's beliefs in the light of today's realities.

7

The
paradox
of
American politics

SEYMOUR MARTIN LIPSET

HE American commonwealth, at
the end of two centuries of independence, abounds with contradictions. Its politics have always been characterized by an extraordinary
emphasis on utopian moralism, which provokes Americans to view
social and political dramas as morality plays within which compromise is virtually unthinkable. As a moralist, consequently, the American tries hard to attain and institutionalize virtue and to destroy
wicked institutions and practices at home and abroad.

Yet, at the same time, the character of our party system suggests
that we are less concerned with using politics for radical social
change than other peoples, that we have a penchant for compromise.
The conservative or pragmatic character of the American polity is
evident in the fact that one of its major parties, the Democratic,
correctly claims to be the oldest party in the world with a continuous
existence, and a line of continuity may also be traced through the
Federalist-Whig-Republican parties. The United States remains the
only democratic country without any socialist representation in its
government.

But once again, contrast these indicators of stability and political
"conservatism" with the relative ease with which a variety of "social
movements," some of which have also been third parties, have arisen

and had a significant impact. The politics of our social *movements* as distinct from that of our *parties* suggests not stability but instability, and emphasizes the power of dissident groups to foster change in America. If we compare the American political system to that of a number of affluent European nations with respect to the frequency and importance of mass movements, the United States would clearly appear to be *less* stable. It does give rise to more mass movements— but these do not result in institutionalized forms of radicalism.

The paradoxical character of American politics is not disclosed by a reading of *The Federalist* or the Constitution. It emerges, rather, from the historical experience of a people which, if not "chosen," is in many respects quite peculiar.

The "dissidence of dissent"

Moralism is an orientation Americans have inherited from their Protestant past. This is the *one* country in the world dominated by the religious traditions of Protestant "dissent"—the Methodists, Baptists, and other sects. The teachings of these denominations called on men to follow their conscience, with an unequivocal emphasis not to be found in those denominations which evolved from state churches (Catholics, Lutherans, Anglicans, and Orthodox Christians). The American Protestant religious ethos has assumed, in practice if not in theology, the perfectibility of man and his obligation to avoid sin, while the churches whose followers predominate in Europe, Canada, and Australia have accepted the inherent weakness of man, his inability to escape sinning and error, and the need for the church to be forgiving and protecting. This basic observation was made by Edmund Burke two centuries ago:

> The People are Protestants; and of that kind most averse to all implicit submission of mind and opinion. . . . Everyone knows that the Roman Catholic religion is at least coeval with most of the governments, where it prevails; that it had generally gone hand in hand with them. . . . The Church of England too was formed from her cradle under the nursing care of regular government. But the dissenting interests have sprung up in direct opposition to all the ordinary powers of the world. All Protestantism, even the most cold and passive, is a sort of dissent. But the religion most prevalent in our northern colonies is a refinement of the principles of resistance; it is the dissidence of dissent, and the Protestantism of the Protestant religion.

The fact of disestablishment—that is, the absence of a state church in America—meant that a new structure of moral authority had to be created to replace the link between Church and State. The withdrawal of government support made the American form of Protestantism unique in the Christian world. Ideological and institutional changes which flowed from the Revolution led to forms of church organization analogous to popularly-based institutions: The United States became the first nation in which religious groups were viewed as purely voluntary organizations, which served to strengthen the introduction of religious morality into politics. Many ministers and laymen consciously recognized that they had to establish voluntary organizations to safeguard morality in a democratic society which lacked an established church. Associations for domestic missionary work, for temperance, for abolition, for widespread education, for peace, for the reduction of the influence of the Masons or the Catholics, and more recently for the elimination of Communists, were organized by people who felt these were the only ways they could preserve and extend a moral society.

The need to assuage a sense of personal responsibility has made Americans particularly inclined to support movements for the elimination of evil—by illegal and even violent means, if necessary. A key element in the conflicts that culminated in the Civil War was the tendency of both sides to view the other as essentially sinful—as an agent of the Devil. And more recently, the resisters to the Vietnam War reenacted a two-century-old American scenario in which a "Protestant" sense of personal responsibility led the intensely committed to violate the "rules of the game." This moralistic tendency in a more secular America has been generalized far beyond its denominational or even specifically religious base. During the (Joe) McCarthy era, a distinguished French Dominican, R. L. Bruckberger, criticized American Catholics for having absorbed the American Protestant view of religion and morality. He noted that American Catholics resemble American Baptists and Presbyterians more than they resemble European or Latin American Catholics: "One often has the impression that American Catholics are more Puritan than anybody else. . . . [An] instance of the same thing was the enthusiasm whipped up by McCarthy among certain American Catholics for 'either virtue or a reign of terror.'" Presumably the emergence of a Catholic left, and the behavior of the Berrigans and others, adds weight to this observation. But agnostic and atheistic reformers in America also tend to be utopian moralists who believe in the perfectibility of man and of civil society and in the immorality, if not specifically sinful character, of the opposition. Like Bruckberger, who reflected a more traditionally Catholic orientation,

Russell Kirk has noted that the purist conservatism of his colleagues on *The National Review* also embodies an American utopian moralistic stance which differs greatly from that of European conservatism, which is much more pessimistic about human nature and society.

Extremism and reform

Focusing on the role and tactics of movements, as distinct from parties, must produce the conclusion that reliance on methods outside of the normal political game has played a major role in affecting change throughout much of American history. While most of the movements have not engaged in violence as such, some of the major changes in American society have been a product of violent tactics resulting from the willingness of those who felt that they had a morally righteous cause to take the law into their own hands in order to advance it. And by extreme actions, whether violent or not, the moralistic radical minorities have often secured the support or aquiescence of some of the more moderate elements, who have come to accept the fact that change is necessary in order to gain a measure of peace and stability. To some extent, also, the extremists on a given side of an issue have lent credence to the arguments presented by the moderates on that issue. Extremists, whether of the right or left, have often helped the moderates in the center to press through reforms.

The most striking example of this sort of behavior in American history was the successful movement to abolish slavery. The radical abolitionists were willing to violate Congressional law and Supreme Court decisions to make their case before the public and to help Negro slaves escape to Canada. Some of them were even ready to fight with arms in order to guarantee that the Western territories would remain free of slavery. John Brown's armed raid on Harper's Ferry played a major role in convincing both Southerners and Northerners that the slavery debate had to be ended either by secession or by some form of emancipation. Conversely, the violence of the first Ku Klux Klan after the Civil War helped convince the North that it had to desist in its effort to prevent white domination of the South. The guerrilla actions of the Klansmen played a major role in re-establishing white Bourbon power and securing the end of Reconstruction.

The women's suffrage movement, as it gained strength, similarly displayed the depth of its commitment by various forms of civil disobedience: illegal demonstrations to disrupt the orderly operation of government, women chaining themselves to buildings, and so forth. Some prohibitionists showed the intensity of their feelings against liquor by ridiculing and ostracizing those who patronized saloons, and on occasion even by violently attempting to prevent dispensers of liquor from doing business.

During the Great Depression, illegal actions were also important. Agrarian movements brought about moratoriums on mortgage-debt collection and changes in various banking laws by their armed actions to prevent the sale of farms for nonpayment of debt. In the cities, the labor movement won its right to collective bargaining in industries that had traditionally opposed it by illegal "sit-ins" in factories in Akron, Detroit, and other places. State governments found themselves helpless to remove workers from factories, and anti-union employers quite often were forced by these actions to accept unions in their plants.

Moralistic politics and movement politics clearly continue today. It is not easy to be a nation which takes morality seriously. Each wave of moralistic protest and reform is necessarily followed by an era of institutionalization in which the inspired utopian hopes become the daily work of bureaucrats, and thus the passion which aroused them is left unsatisfied. Further, it becomes obvious that the problems are much more complex than assumed in the simple solutions proposed by protest movements, whether for emancipation, civil service reform, prohibition, conservation, women's suffrage, control of trusts and monopolies, intervention for economic and welfare purposes, or war against reactionary foreign states. What is often worse is the realization that what Robert Merton has termed the "unintended consequences of purposive social action" have brought new evils.

And even when the reforms accomplish their manifest purpose, they still leave a society which is highly immoral from the vantage point of those who take seriously the constantly redefined and enlarged ideals of equality, democracy, and liberty. Corruption is perhaps endemic in a competitive meritocratic society, whether capitalist or Communist, and periods of prolonged prosperity in which many become visibly wealthy generally witness the spread and institutionalization of corruption. Privilege, too, always seeks to entrench itself, and is generally able to do so in such times. Thus it is not surprising that new generations of Americans recurrently respond to some event or crisis which points anew to the gap between the American ideal and reality by supporting a new protest wave premised on the assumption that a corrupt and morally sick America must be drastically reformed.

The United States has been in such a period since the mid-1960's. As in the past, the moralistic reformers have thrown a wide net of criticism over American institutions and behavior. Not surprisingly, those most motivated to support these criticisms place them in a traditional Protestant context—of good versus evil, God versus Satan, progress against reaction—and define American society as totally evil in much the same terms as abolitionist William Lloyd Garrison did when he tore up the Constitution as a compact with the Devil. The rhetoric of American politics, as Will Herberg noted many years ago, normally goes far beyond the substantive content of the issues: Frankin Roosevelt denounced his opponents as "Copperheads"—i.e., as the equivalent of traitors to the Northern cause during the Civil War—while his rivals identified Roosevelt's New Deal as Communist-inspired.

Moralism and foreign policy

The strength of moralistic pressures may be seen most strikingly in reactions to foreign policy issues. There have been three uniquely American stances: conscientious objection to unjust wars, non-recognition of "evil" foreign regimes, and the insistence that wars must end with the "unconditional surrender" of the Satanic enemy. Linked to Protestant sectarianism, conscientious objection to military service was until recently largely an American phenomenon.[1] To decry wars, to refuse to go, is at least as American as apple pie. Sol Tax of the University of Chicago, who attempted to compare the extent of anti-war activity throughout American history, concluded that as of 1968 the Vietnam war rated as our *fourth* "least popular" conflict with a foreign enemy. Widespread opposition has existed to all American wars, with the possible exception of World War II, which began with a direct attack on United States soil. Large numbers refused to go along with the War of 1812, the Mexican War, the Civil War, and the Korean War. They took it as self-evident that they must obey their conscience rather than the dictates of their country's rulers.

The supporters of American wars invariably see them as moralistic crusades—to eliminate monarchical rule (the War of 1812), to defeat

---

[1] It could also be found on a less widespread scale in other English-speaking countries, where, however, it has been less prevalent since a much smaller proportion adhere to the "dissenting" sects.

the Catholic forces of superstition (the Mexican War), to end
slavery (the Civil War), to end colonialism in the Americas (the
Spanish-American War), to make the world safe for democracy
(World War I), and to resist totalitarian expansionism (World War
II, Korea, and Vietnam).[2] Unlike other countries, we rarely see ourselves as merely defending our national interests. Since each war
is a battle of good versus evil, the only acceptable outcome is "unconditional surrender" by the enemy.

George Kennan has written perceptively of the negative consequences of the "carrying-over into the affairs of states of the concepts
of right and wrong." As he notes, when moralistic "indignation spills
over into military context, it knows no bounds short of the reduction
of the law-breaker to the point of complete submissiveness—namely,
unconditional surrender." Ironically, a moralist "approach to world
affairs, rooted as it unquestionably is in a desire to do away with
war and violence, makes violence more enduring, more terrible, and
more destructive to political stability than motives of national interest. A war fought in the name of high moral principle finds no
early end short of some form of total domination."

The stalemated struggle with Communism is, of course, a blow to
this sense of a moralistic contest which must end with the defeat
of Satan. America's initial reaction to Communism was one which
implied no compromise. After each major Communist triumph—
Russia, China, Cuba—we went through a period of refusing to "recognize" this unforgivable, hopefully temporary, victory of Satan.
(This behavior contrasts with that of Anglican conservatives such
as Churchill, or Catholic rightists such as DeGaulle or Franco, whose
anti-Communism did not require "non-recognition.") Ultimately, the
facts of power, and in the 1930's the rise of another even more bel-
ligerent enemy, Nazism, pressured the United States to deal with
Communism. During World War II, we were even forced to ally
ourselves with the Soviet Union and Communist partisan movements. Our initial reaction to this necessity is indicative of the way
in which moralism affects the national purpose: The Soviet Union
was quickly transformed into a beneficent, almost democratic state.
Eddy Rickenbacker wrote in glowing terms in the *Reader's Digest*
that the Soviet Union had practically become a capitalist society.
Both Stalin and Tito were presented as "progressive" national leaders and heroes comparable to classic American figures. Franklin
Roosevelt, for a time, allowed himself to see Stalin as a leader of

---

[2] Both abolitionists and others objected to fighting Mexico, and the Mexican
army actually formed units manned by deserters from the United States forces.

anti-imperialist forces with whom the United States could cooperate in planning the post-war world, even against the French and British imperialists.

Subsequent Soviet behavior in Eastern Europe and Berlin during the latter years of the war and the early post-war years destroyed this effort to transform the image of Communism. The Communist victory in China, reinforced by the events of the Korean War, produced a reaction comparable to that directed against the Soviets after 1917. The United States refused to recognize evil in China (and later in Cuba). It engaged in an internal heresy hunt seeking to find and eliminate the traitors at home responsible for the inefficacy of our anti-Communist efforts abroad. The McCarthyite period, of course, coincided with a hot war against Communism, the Korean War, and thus also represented an effort to repress critics of the war, corresponding to previous waves of wartime repression.

The reaction to the Vietnamese War also reveals the extent to which the need for moralistic politics, particularly in foreign policy and wartime, affects the behavior of Americans. For this conflict was the first war waged by the American political elite which did not include a moralistic crusade designed to gain total victory. From the start, an unwillingness to get involved in a major war in Asia, a desire to avoid provoking Chinese and/or Soviet direct military intervention, and—not least—a fear that an anti-Communist crusade would reawaken a right-wing McCarthyist reaction led John F. Kennedy and Lyndon Johnson to underplay deliberately the anti-Communist ideological crusade as a justification for the war. It was defined as a limited war in which the United States would do as little as was required to prevent North Vietnam from taking over the South. There was almost no government-inspired propaganda designed to portray the repressive character of the North Vietnamese state.[3] Pro-war journalists, seeking information about Communist atrocities or pictures such as those of the heads of village leaders on spikes, were actually denied them by the Defense Department to avoid inflaming public opinion. For some years after the anti-war movement reached mass proportions, Lyndon Johnson was quoted by intimates as worrying much more about right-wing "hawkish" opposition than that arising on his left.

---

[3] French expert on Vietnam Jean LaCouture described North Vietnam as the most Stalinist regime in the Communist world. North Vietnam continued to include Stalin in its pantheon, and was led by a die-hard Stalinist, Ho Chi Minh, who had been a major Comintern representative in internecine Communist battles for many decades, and who engaged in a bloodbath against Trotskyists and other radicals after taking power in Hanoi.

Given the unwillingness of the government to motivate a crusading atmosphere, the obvious breakdown in the monolithic image of a totalitarian Communist empire in the wake of de-Stalinization in the Soviet Union—revolts and nationalist regimes in eastern Europe and the Sino-Soviet split—and, finally, the inability of hundreds of thousands of U.S. troops to defeat the Viet Cong and the North Vietnamese, the rise of a moralistic mass opposition to the war was inevitable. An unengaged morality shifted to the side of the anti-war forces, and as the war was prolonged, it was inevitable that they would win.

The dilemmas of détente

Whether the failure of the spirit of a moralistic crusade to emerge during the Vietnam war represents a basic change in popular thinking is, of course, difficult to answer. It may be that efforts to appeal to a purist patriotism are no longer possible, that acceptance of simplistic moralism is much reduced by the greater awareness of the complexities of modern politics which results from higher education and the more sophisticated mass media. Although I doubt that such a basic change has occurred, it is true that the proliferation of nuclear weaponry—of atomic deterrents—means that future wars cannot be fought through to "unconditional surrender." Hence the only alternatives are cold wars, limited wars, and détente. In a situation in which total war means mutual annihilation, both policy makers and the public alike are pressed to define international conflict in non-moralistic terms and to seek evidence that the other side is not impossibly evil. The absence of significant public support for the moralistic right-wing Goldwater-Reagan program for total war in Vietnam, as compared to that which rallied around Joe McCarthy during the Korean War, attests to a major change in attitudes.

The same dilemma posed by Vietnam is being replayed in a less sanguine context in current United States détente policy. The Nixon-Ford-Kissinger policy which seeks détente with China and the Soviet Union inevitably undermines efforts to sustain strong American defense and foreign-alliances programs, as well as intelligence activities premised on the need to resist expansionist Communist systems and movements.

The current situation is but the most recent illustration of the well known political principle, "You cannot have your cake and eat it too." To the moralistic posture Communism today is or is not evil. It is or is not a threat to democracy at home and abroad. If it is, it must be resisted—a feeling many Americans still have. If it is not, then it makes little sense to assign a major share of our limited resources to anti-Communist defense and alliance efforts. And in this context, many non-Communist Americans "understand" a strong defense program as being little more than self-interested efforts by those who seemingly gain from such a policy, the "military-industrial complex."

This is, of course, a vastly oversimplified description of the nature and sources of the American reaction to Vietnam and détente. Obviously, from the perspective of policy-makers in the White House and in the State and Defense departments, United States intervention in Vietnam was in the national interest. The issue of American credibility in supporting non-Communist SEATO allies concerned President Kennedy at the beginning of American involvement and continued to affect President Ford and Secretary Kissinger at the end. Once direct involvement had begun, President after President indicated that he did not want to be the first one to preside over an American defeat. Henry Kissinger seemingly was affected by the Weimar analogy, by the consequences of an American defeat for the legitimacy of the regime, particularly with regard to the loyalty of the military.

The destructive effects of the failure in Vietnam on the American sense of its world responsibility are reinforced by the concomitant rejection of American AID and alliance policies by much of the Third World and many Europeans. Regardless of the economic consequences of such efforts, most Americans supported these in the spirit of missionary efforts to improve and uplift the way of life of oppressed foreigners. Rejection by those we are trying to help is a blow to this self-conception. As Gabriel Almond has emphasized in his analysis of *The American People and Foreign Policy*, "If generous actions, motivated by moral and humanitarian considerations, are accepted without gratitude, are misinterpreted, or are unrequited, a 'cynical' rejection of humanitarianism may follow, resulting from the humiliation of having been 'played for a sucker.'"

The depression following the failure of a supposedly humanitarian foreign policy after World War I resulted in a wave of isolationism and cynicism about the motives of other peoples and their leaders. We seemingly are in a comparable period today, as reflected in the increasingly popular unwillingness to continue our military support for Korea. As of December 1974, the Harris Survey indicated that only 14 per cent of the American people favored using American military forces to defend South Korea against a Communist attack.

Moralism and domestic policy

An even more striking consequence, however, of the loss of the moralistic self-image has been the drastic decline in confidence in American domestic institutions, in the American way of life itself. Ten years ago, Yale political scientist Robert Lane reviewed the comprehensive evidence from a large number of opinion surveys with respect to attitudes toward the American system, government, and political leadership. On every indicator, the trends revealed sizable majorities—often over three quarters—giving positive responses. As we all know, with the benefit of hindsight, Lane's article appeared just as the era he described as an "age of consensus" was ending. From 1965 on, the answers of the American public and leadership groups to comparable questions went steadily downwards until, by 1974, only 30 per cent or so were responding positively to the questions about the system to which over 70 per cent had been favorable a decade earlier.

These changes in attitudes go beyond an increased distrust of government and of politicians. The Harris survey, which has repeated the same questions year after year, reports that the percentage agreeing with the statement, "What you think doesn't count anymore," rose steadily from 37 per cent in 1966 to 61 per cent in 1973. Those accepting the proposition, "The rich get richer and the poor get poorer," grew from 45 per cent in 1966 to 76 per cent in 1973. Even reactions to the broad non-political view that you "feel left out of things going on around you," changed from nine per cent agreement in 1966 to 29 per cent in 1973. On a four-question scale of "alienation, powerlessness and cynicism," those giving disenchanted responses rose from 29 per cent in 1966 to 55 per cent in 1973.

Although the events stemming from Vietnam and the moralistic protest it sustained clearly were the catalytic events which stimulated the crisis of confidence and the growth of alienation, they were, of course, strongly reinforced by the Watergate exposés, and more recently by the recession. Harris reports that when a sample of the population were asked in the fall of 1973 "to explain their concern, one quarter of those sampled—the largest segment—volunteered the belief that 'government leaders are corrupt and immoral.'"

These changes in popular feeling about basic institutions and leaders reflect a declining self-confidence by many Americans that their country still is the *good society,* that it is on God's side. For the deviations in behavior which accompanied protests against the Vietnam war involved acceptance of practices, once totally repugnant to the traditional morality—drugs, easy sex, abortion, pornography, no prayers in schools, less stringent punishments for severe crimes, etc.—which in turn engendered the feeling in many Americans that their own children were not acting morally. For those who still wanted to believe in the Manifest Destiny of America, there was the evidence that anti-Communism, non-recognition and boycotts of the enemy, and support for anti-Communist forces simply did not work in dealing with China, Cuba, and North Vietnam.

Yet from another perspective, these reactions—the decline in confidence in institutions and leaders—may be perceived as the other side of moralism, much as the historic American penchant for anti-war activity is the reverse image of a belief in Manifest Destiny. As a myriad of articles on foreign reaction to Watergate and CIA revelations have emphasized, Europeans, Israelis, Japanese, and others have frequently expressed bewilderment at what looks to them to be an "over-reaction" of Americans to types of government behavior which foreigners take for granted as inherent in the relation of politics to society. Here is another illustration of the differences in outlook between societies which assume some natural wickedness of men, and one whose ethos is premised on a utopian, religious belief in human perfectibility. Americans retain both the capacity to be shocked by evil and the motivation to resist it.

Conspiracy-theory logic

The Watergate events, both the background of the happening as well as the reaction to the exposé, may also be seen as the outcome of moralistic emphases in politics. As Earl Raab and I have emphasized in our historical analyses of right-wing movements in America (*The Politics of Unreason, 1790-1970*), these movements—the anti-Illuminati agitation of the 1790's through the Anti-Masons, assorted nativist crusades for over a century, the Klan of the 1920's, McCarthyism, the Birch Society, and the underlying ideological syndrome dominant in the Nixon White House—assumed the necessity to combat the secret internal enemies of the United States, who were linked to immoral forces abroad. After taking office in 1969, many in the White House linked the violent "extra-parliamentary" agitation of the New Left and the anti-war movement with foreign support. When the FBI and the CIA were unable to locate evidence of foreign involvement, White House staffers concluded that a new intelligence operation, under White House control, was needed.

Conspiracy-theory logic, fostered by moralistic conceptions, enveloped the White House. Much as extremists, both of the left and of the right, have found it impossible to accept the fact that Lee Harvey Oswald was a loner linked neither to the CIA nor to Castro, so the White House could not accept a non-conspiratorial interpretation of Daniel Ellsberg's turnabout. Faced with the "betrayal" of a once-trusted supporter of the Vietnam war, they refused to see Ellsberg as a man who, like many Americans, has always shown the need for passionate commitment—who, in the words of the CIA psychological profile which they rejected, has always been "either strongly for something or strongly against." In the view of the White House patriotic moralists, Ellsberg *had* to be part of a broader conspiracy, and anyone who doubted this—even the CIA or J. Edgar Hoover—must have been either misled or somehow corrupted.

In short, the views and behavior which led to Watergate were typical, at least in form, of American backlash moralist extremism. Those who hired the "plumbers" and then shifted their arena of operation to CREEP were, like their opponents in the anti-war movement, convinced that *their* moral objectives justified the use of illegal means to destroy evil forces. They behaved like those involved in earlier moralistic social movements who sought to attain their ends regardless of the damage caused by their tactics and rhetoric to the society. The moralists typically react in horror to the corrupt and illegal—sometimes extreme—tactics of their opponents, unaware that they themselves are engaging in similar illegal behavior. The struggle between the anti-war movements and the administrations during the Indochinese War is a good illustration of the interaction of moralism and use of illegal methods on both sides. Knowledge of what the one side was doing only intensified the justification offered by the other side for the use of extreme tactics. Archibald Cox, in discussing the events which led up to Watergate, has stressed the extent to which outrage against the illegal acts and terrorism employed by the anti-war forces justified extraordinary measures in the minds of many in the Nixon entourage. From the White House, to the opponents of the war, to Judge Sirica (who used the tactic of threatening little men with extraordinary sentences to bully them into talking), self-righteous moralists felt sanctimonious is using violent, illegal, or cruel methods to defeat their "enemies."

Compromise and coalition

But the American propensity to moralistic and extreme politics has been effectively countered by the compromise coalition politics characteristic of our party system. The need to construct coalitions, often involving groups with sharply opposed interests and values, and the development of party ideologies and rhetoric appropriate to keeping such coalitions together repeatedly undercuts the moralistic passion of social movements. If the country is to be governed, party leaders must find ways of encapsulating and incorporating indignation.

In a two-party system, both parties aim at securing a majority. Elections become occasions for the two parties to seek the broadest base of support by convincing divergent groups of their common interests. In contrast to much of Europe, where divergent *parties* join coalitions *after* elections to form cabinets, in the United States divergent *factions* come together *before* elections to elect Presidents or governors. The system thus encourages compromise and the incorporation into party values of those general elements of consensus upon which the polity rests. The normal lack of emphasis on ideological differences that is inherent in two-party systems has the further consequences of reducing intense concern with particular issues and sharpening the focus on party leaders.

The structural factor which has made this balancing act possible in America is the electoral system which is almost required by the Constitutional separation of executive and legislative powers. The Presidential system, almost unique in the democratic world until DeGaulle attempted to imitate it in France, dictates a two-party system. Unlike the situation in parliamentary countries, where new and/or small, ideologically committed parties may establish a core of elected representatives—easier if a system of proportional representation is used, more difficult but possible with single-member districts if the parties have local strength in distinct regions or political units—the effective constituency for electing an American President is the entire country, and for a governor the whole state. This has had the effect of preventing American third parties from building up local constituency strength as labor, agrarian, religious, or ethnic parties have done in most single-member constituency systems. Evidence for this interpretation is the fact that third parties in the United States have gained their greatest strength in municipal, and occasionally, state contests, and in Congressional elections held in non-Presidential election years, but have almost invariably lost strength in subsequent Presidential elections. Essentially, the division into two parties is maintained by those factors which lead people to see a third-party vote as "wasted." Thus, since opinion surveys began in the 1930's, they have indicated a steady decline in support, as Election Day nears, for third-party candidates such as William Lemke in 1936, Henry Wallace in 1948, and George Wallace in 1968.

Another factor which contributes to the stability of the American two-party system is the decentralized structure of the major parties, flowing both from the federal system and the separation of powers. Since a parliamentary defeat for the majority party does not affect the executive's tenure in office as it does in other countries, party discipline in Congress is weak, and allows for cross-party alignments on particular issues. Congressmen who represent minority orientations within one party can vote against the party leadership and the President in order to please their constituents. Further, primary elections give minority interests an opportunity to openly express themselves in opposition to those in office in their own party, and thus help to keep these minorities within the party.

As House majority leader "Tip" O'Neill noted recently, "If this country were any other, it would have a highly divisive multi-party system." Such a system probably would include such diverse elements as the following: 1) a labor or socialist party based on urban workers and perhaps on ethnic minorities outside the South; 2) a Northern conservative party, supported by the urban middle class; 3) a Southern conservative party backed by the more privileged elements; 4) a Southern populist party; and 5) a declining farmer's party based on elements outside the South. Groups with a strong sense of solidarity and interests living in ethnographically distinct areas—such as the blacks, Catholics, and fundamentalists—might also have formed their own parties.

Moralistic backlash

The construction of electoral coalitions on a national level is an extremely delicate task, and often requires that party leaders deemphasize ideological (moralistic) appeals and focus on either "throwing the rascals out" or keeping them out. Critical change periods—e.g., the slavery issue and the Civil War, the rise of agrarian populism in tandem with the industrialization at the end of the 19th century, the Great Depression—have witnessed realignments of party supporters into new coalition systems. And such periods of realignment have usually produced coalitions that have persisted for some decades. They break down, however, as groups find themselves out of step with their party on specific new issues, while the alternative party is not prepared to make their causes its own.

Thus, extra-party "movements" arise to press for moralistic causes, which are initially not electorally palatable. These extra-major-party movements have taken various forms, most often emphasizing a single special issue, but sometimes cohering around a broader ideology. Among their central themes have been anti-Masonry, nativism (particularly focusing around the sins of Catholics), abolitionism, feminism, prohibition, peace, socialism, clean government, conservation, and the environment.

Such movements are not doomed to isolation and inefficacy. If mainstream political leaders recognize that a significant segment of the electorate feels alienated from the body politic, they will readapt one of the major party coalitions. But in so doing, they temper much of the extremist moralistic fervor. Sometimes this may be done by accommodations in rhetoric, but the results are often actual changes in policy. The protestors are absorbed into a major party coalition but, like the abolitionists who joined the Republicans, the Populists who merged with the Democrats, or the radicals who backed the New Deal, they contribute to the policy orientation of the newly formed coalition.

Among the many moralistic strands that have affected the American polity for well over a century, two stand out: One grew out of intergroup tensions in multi-ethnic America, which were reflected in a variety of mass-based movements dedicated to nativist, religious, or racist intolerance; the other was the outcome of a persistent strain between the values of segments of the highly-educated intellectual elite and those of a materialistic economic system. The first, though massively present in such earlier movements as the Anti-Masons (1828-1832), the Native Americans (1840's), the Know-Nothings (1852-1857), and the American Protective Association (1890-1897), and continuing as a major tendency today, had its most important impact in the 1920's. Movements stemming from the intelligentsia, on the other hand, reached their high point of influence in the late 1960's but may be even more important in the future.

The decade of the 1920's witnessed political repression (the Red Scare, the Palmer Raids), moralistic intolerance (prohibition, the anti-evolution teaching laws), and racial and religious bigotry (the rise of the multi-million member Ku Klux Klan, the diffusion of Henry Ford's mass circulation, anti-Semitic paper, the *Dearborn Independent*, and the passage of racist-inspired immigration legislation). Henry Ford, viewed as a possible Presidential candidate, led in opinion polls in 1923 and was offered third-party nominations. Defenders of the Klan had close to half the delegates at both the Democratic and Republican Conventions in 1924.

This massive protest wave may be seen as a backlash reaction
to the fact that as a result of broad structural changes American
society was becoming cosmopolitan, secular, and metropolitan—with
negative consequences for evangelical values. Traditional Protestantism was well on its way to becoming a minority culture, given
the growth of the disproportionately Catholic, Jewish, and secularist
cities as centers of communications and power. As historian Arnold
Rice noted: "The 1920's meant modernism. And 'modernism' among
other things meant the waning of church influence ... the discarding
of the old-fashioned moral code in favor of a freer or 'looser' personal
one, which manifested itself in such activities as purchasing and
drinking contraband liquor, participating in ultra-frank conversa-
tions between the sexes, wearing skirts close to the knees, engaging
in various extreme forms of dancing . . . and petting in parked cars."
As one Klan leader put it poignantly, "We have become strangers
in the land of our fathers."

In reaction to this growth of "modernism," the 1920's produced a
Dixiecrat-Republican coalition, in which the evangelical Protestant
non-metropolitan Republicans of the North joined with the Protestant Democrats in the South against the big-city Catholic Democrats.
And it was this Dixiecrat-Republican coalition that put across measures like prohibition and immigration restriction. Given the identification of the northern Republican party with these issues, it is
not surprising that the G.O.P. gained considerably in Presidential
voting. The rise of the Klan, the appeal of Henry Ford, and the massive increase in Republican support each reflected in a different way
the desire of many Americans to restore an America which had been
changed by war, urbanization, and heavy waves of non-Nordic immigration to reflect once more the values of a rural, moralistic Protestant society. Conversely, emerging Catholic, Jewish, and, ultimately, black Americans were drawn to the Democratic party coalition, which in the 1930's became a majority when the Depression
and the growth of organized labor gave it significant strength among
urban Protestants, particularly the trade unionists among them.

Nativism and the new ethnicity

Backlash politics, however, was to continue to provide the basis
for new forms of nativism which upset party loyalties. Ironically,
nativism began to attract support from the ranks of the more recently established Catholic population—who themselves had been
the chief victims of such moralistic sentiments before 1930. As the

blame for the moral threat to American values shifted from persons
of alien *origin* and *religion* to persons of alien *ideas* and *values,*
particularly Communists, non-WASP elements could become partisans of the new nativism. Thus, segments of the Catholic population briefly broke with the Democratic coalition to support two
rightist Catholic spokesmen, Father Charles Coughlin in the 1930's,
and Senator Joseph McCarthy in the early 1950's. Both had a clear
impact on public policy. Coughlin (and Huey Long) ironically
helped push Franklin Roosevelt to the left on economic issues by
emphasizing the plight of the poor and the need for government action. McCarthy contributed to the increased intensity of the United
States cold war posture.

In a curious way, developments in recent decades comparable to
those which produced the backlash movements of the 1920's have
led members of some of the very groups whose presence and growing influence inspired racist reaction in the earlier period to engage
in comparable behavior today. A combination of the "invasion" of
the central cities by blacks and Hispanic Americans and the cultural
changes in sexual morality, patriotism, religious beliefs, law and
order, and the nature of education revived backlash cultural and
ethnic politics. Many lower-middle-class and working-class whites
began to feel that a large segment of the Democratic party leadership, of the newly-emboldened liberal Protestant and Catholic
clergy, and of the WASP and Jewish "suburban elites" had lost
interest in them, that they were concerned solely with the plight of
the blacks, middle-class women, and with "permissive" changes in
the cultural arena.

The reaction has taken two major forms, a revitalization of organized ethnicity, largely a class response by white workers to their
seemingly having been abandoned by political institutions, and massive support given in the polls to George Wallace, who has exploited
the sense of isolation of the "forgotten working man." Wallace's
backing in opinion surveys and in primary contests points up the
vast numbers who currently feel excluded from both major party
coalitions, one of which they identify with the blacks and cultural
license, and the other which they see linked to big business and un-
interested in their economic class interests.

Left-wing moralism

The adequacy of the two party system to produce effective coalition government has also been undermined in the past 10 years

by another element in the polity: the well-educated, affluent, cosmopolitan elites who have become a major force in contemporary America. This is clearly not a new tendency. Segments of these strata have sought to "clean up" and to purify America for almost as long as the nativists, and have been repeatedly frustrated by the non-moralistic, compromise character of the two-party system. Historians have traced a pattern of continuity in the leadership and activist core of "anti-materialistic" reform movements from the pre-Civil War abolitionists, through the clean government "mugwump" (anti-party) movement of the late 19th century, to the anti-machine and anti-trust Progressives of the pre-World War I era, down to the assorted segments of the protest "movement" of the 1960's and early 1970's, which have emphasized clean environments, clean government, clean business, honesty in advertising, and equal rights for oppressed minorities, opposed America's effort to impose its political will abroad, and sought to eliminate the remnants of ascetic Protestant morality concerning sex and the behavior of youth.

Writing in 1873 about the situation before the Civil War, Whitelaw Reid, abolitionist activist and editor of the New York *Tribune,* noted that "exceptional influences eliminated, the scholar is pretty sure to be opposed to the established. . . . As our politics settled into the conservative tack in the pre-Civil War decades a fresh wind began to blow about the college seats, and literary men, at last, furnished inspiration for the splendid movement that swept slavery from the statute book. . . ." And commenting on the rise of "mugwump" opposition to the party system in the 1880's, James Bryce described the role of intellectuals, academics, and their college-educated followers in terms not dissimilar from those used to analyze their impact on the massive protest wave of the 1960's:

> The influence of literary men [on politics] is more felt through magazines than through books. . . . That of the teacher tells primarily on their pupils and indirectly on the circles to which those pupils belong, or in which they work when they have left college. One is amused by the bitterness—affected scorn trying to disguise real fear—with which "college professors" are denounced by professional politicians as unpractical, visionary, pharisaical, "kid-gloved," "high-toned," "un-American," the fact being that a considerable impulse towards the improvement of party methods, towards civil service reform, and towards tariff reform, has come from the universities, and has been felt in the increased political activity of the better educated youth. The new generation of lawyers, clergymen and journalists . . . [has] been inspired by the universities, particularly of course by the older and more highly developed institutions of the Eastern States with a more serious and earnest view of politics. . . . Their horizon has been enlarged, their patriotism tem-

pered by a sense of national shortcomings, and quickened by a higher ideal of national well-being. The confidence that all other prosperity will accompany material prosperity, the belief that good instincts are enough to guide nations through practical difficulties—errors which led astray so many worthy people in the last generation, are being dispelled. . . . The seats of learning and education are at present among the most potent forces making for progress . . . in the United States.

Similar views were expressed concerning the role of the academy in fostering opposition to the Spanish-American War and to its aftermath, the struggle against Filipino guerrillas. A historian of academe, Laurence Veysey, notes that "faculty opposition to imperialism during the 1890's was observed as general all over the country." While the short war against Spain was still on, Oliver Wendell Holmes, Jr., commented to a friend, "I confess to pleasure in hearing some rattling jingo talk after the self-righteous and preaching discourse which has prevailed to some extent at Harvard College and elsewhere." An article in the *Atlantic Monthly* in 1902 reported that college professors had acquired a reputation for taking obstructionist political positions: "Within a twelve-month college teachers have been openly denounced as 'traitors' for advocating self-government for Filipinos. In many a pulpit and newspaper office . . . it was declared that the utterances of college professors were largely responsible for the assassination of President McKinley."

The intellectual elite

What Lionel Trilling has called the "adversary culture" represents the general contempt which the creators of culture, and their fellow-travelers, have felt against the materialist philistines concerned with bourgeois values. Such an orientation has led intellectuals in different societies to foster diverse anti-materialistic, anti-bourgeois, and anti-technological ideals, sometimes linked to Catholic medieval conservatism, to fascism, and of course in recent times to forms of socialism, communism, and left-wing nationalism.

In the United States the characteristic stance of the intellectuals, and the educated strata which support them, has been moralism. They have scorned society for failing to fulfill agreed-upon liberal values. They have repeatedly challenged those running the nation with the crime of heresy, with betraying the American Creed. In so doing, however, the educated critics in America have focused on cultural critiques, including a desire for a "clean" reformed society, rather than on radical institutional change. As a group, however, they tend to be fascinated with power—exhibited, at times, in exaggerated fear of it when seemingly directed against them, and at other times in adoration of a charismatic leader with whom they can identify. This latter propensity has led them repeatedly back into two-party coalition politics.

Thus before World War I, many coming from Ivy League and "secure upper-middle-class" families appeared to reject the political system, backing the egalitarian and seemingly anti-business objectives of the Progressives and Socialists. But this "movement" of radical culture-critics broke down when faced with the seduction of "intellectual power" in the form of Wilson's administration. Again, during the 1920's many of the most outspoken American intellectuals expressed renewed antagonism to the national system, and its dominant business class from which came (in the words of Frederic Hoffman) "dullness, stupidity, aggressiveness in commerce, conformity to the remnants of traditional morality, and a moral opportunism linked with certain blind convictions about the economic status quo." The revulsion soon turned intellectuals to the left. With the beginning of the Great Depression, large segments of the intellectual and educated communities flocked to support the Communist and Socialist parties, and assorted protest movements designed to realign American politics. Yet, as in the time of Wilson's New Freedom, a reform President of aristocratic origins, Franklin Roosevelt, who openly flattered intellectuals and incorporated some of them into his administration as experts, was able to win their enthusiastic support—particularly as they were able to combine interaction with "anti-business" power in America and a pseudo-radical love affair with the Soviet Union and the newly Americanized Communist party, which had become a member of the Democratic party coalition.

More recently, Daniel Bell, John Kenneth Galbraith, and others have identified the intensified strength of this leftward tendency as a reflection of structural changes in advanced or "post-industrial" society, which have created a massive intelligentsia, a highly educated oppositionist class. At their core is the "educational and scientific estate," including hundreds of thousands of professors—some employed in research and development positions outside the academy, some in the media, and some in welfare, governmental and other nonprofit institutions—and the 10 million college students.

Following World War II, the intellectual elite shifted the focus of its moralistic critique from political and economic institutions to a corrupted culture and environment. It found its prototypical leader in Adlai Stevenson, a man who deemphasized economic and class issues in favor of a stress on the decline of moral, cultural, and ecological standards. Many sought to reform and clean up American politics, and Democratic reform movements, which centered in the west side of Los Angeles, Hyde Park in Chicago, Cambridge, and the west side of Manhattan, attracted considerable support among the growing intellectualized professional strata. The linkage of cultural-academic concerns to party politics in the 1950's made possible the intense politicization of the intellectuals in the ensuing decade. But it is interesting to note that the cultural-political folk hero of that subsequent period, John F. Kennedy, initially was not popular among intellectuals when he ran for President in 1960: They were repelled since his record revealed no great political passions and he had sat out the fight against McCarthy, while other members of his family, including his brother Robert, had actively supported McCarthy. Stevenson and, to a lesser extent, the erstwhile academic, Hubert Humphrey, were the preferred candidates of the politicized intellectuals.

"Movement" politics

During the mid-1960's, American intellectuals once again assumed the role of "moralists" with respect to political criticism, denouncing the system for betraying its own basic democratic and anti-imperialist beliefs. Beginning with the faculty-initiated teach-ins against the Vietnam war in 1965, intellectuals played a major role in sustaining a mass anti-war movement out of which a large radical constituency emerged. Given the identification of both major parties with the war, the anti-war agitation contributed to a decline in partisan identification, and added to the growing number of "independents" on the left, much as Wallace-linked sentiments produced large numbers of independents on the right. There can be little doubt of the efficacy of this protest movement of the intelligentsia. A variety of statistical data validates Galbraith's 1971 boast that "It was the universities . . . which led the opposition to the Vietnam War, which forced the retirement of President Johnson, which are leading the battle against the great corporations on the issue of pollution, and which at the last Congressional elections retired a score or more of the more egregious time-servers, military sycophants, and hawks." The "movement" was, of course, able to impose its will on the Democratic party nomination in 1972, but in the process it drove out of the party a sizeable segment of the less affluent, many of whom had voted for George Wallace in the primaries.

In the mid-1970's the American political scene appears to be in total confusion. Inchoate movements or tendencies rather than parties dominate the landscape. Temporary coalitions form, dissolve, and reform around municipal, state, and national politicians. The Republican party remains controlled on the local levels by moralistic ideologues such as Barry Goldwater and Ronald Reagan—who support economic laissez-faire and cultural conservatism—a combination with limited public support. Men like Frank Rizzo, Richard Daley, and George Wallace appeal to the less affluent and less educated whose traditional political home, since the early New Deal days, has been the Democratic Party. And the growing stratum of intelligentsia, produced by the massive system of higher education (now incorporating 600,000 faculty and 10 million students) and supported by the media elite, also feels alienated from a political and economic system which it sees as repressive and corrupt.

Each major strand is looking for a political home which will give expression to its particular form of moralistic politics. The air rings with code words like busing, crime in the cities, welfare frauds, Watergate, Secret Police repression, business corruption, ecology, and the like. But the very intensity of these beliefs makes difficult the kind of compromise which has sustained the two-party system.

The breakdown of parties and the political disorganization which appears to characterize the contemporary scene are far from unique. In the past, dislocations of the great party coalitions have also been accompanied by the process of "polarization." This term generally describes the condition whereby significant sections of the population move to the left and right of normal two-party politics. In the 1820's and 1830's the rise of the Anti-Masons was paralleled by the Workingmen's parties; the nativists of the 1840's and 1850's by the Free Soilers, Liberty Party, and Abolitionists; the anti-Catholic American Protective Association of the 1890's by the Populists; the massive Klan of the 1920's by the Progressive and Farmer-Labor movements; the Coughlinites of the 1930's by significant, active, leftist radical movements; and in the last decade, George Wallace and his followers by the "movement" encompassing the "New Left," the "New Politics," the "Black Revolution," and the opposition intelligentsia.

This polarization process always involves two forces which react not only to specific issues but to each other. And as this occurs, politics increasingly comes to be perceived in purely moralistic terms, as involving a struggle between good and evil forces rather than as a series of collective bargaining issues.

The increased commitment to extra-partisan moralism in the 1960's produced the most serious breakdown of political restraint since the early 1920's. Civil disobedience often verging into violence has become an accepted tactic among leftist militants, civil rights advocates, and opponents of school busing alike. Governor Wallace, the editors of the *New York Review of Books,* and John Ehrlichman all justified illegal acts that were intended to defeat enemies of the republic.

Freedom, the underlying principle of a democratic society, requires a commitment to restraint, a willingness not to do anything to undermine the basic set of conventions which enable men of different values and interests to live together. In an effort to avoid or to end the law of the jungle, men set up constitutional and legal curbs on what they may do to one another to attain desired ends. It is no accident that the Bill of Rights is worded not positively, but largely in the language of restraint: "Congress shall make no law. . . ."

Yet the chief bulwark against a breakdown in restraint is not the Constitution as such but the American two-party system. The current president of the American Political Science Association, James MacGregor Burns, has described well the role of the party system:

> Majority rule in a big, diverse nation must be moderate. No majority party can cater to the demands of any extremist group because to do so would antagonize the great "middle groups" that hold the political balance of power and hence could rob the governing party of its majority at the next election. A democratic people embodies its own safeguards in the form of social checks and balances—the great variety of sections and groups and classes and opinions stitched into the fabric of society and thus into the majority's coalition. . . . Moreover, the majority party—and the opposition that hopes to supplant it—must be competitive; if either one forsakes victory in order to stick to principle, as the Federalists did after the turn of the century, it threatens the whole mechanism of majority rule. Majoritarian strategy assumes that in the end politicians will rise above principle in order to win an election.

The two-party system has served to moderate the moralistic passions that are inherent in what Lincoln called the American "political religion." That system, however, is finding this task of moderation increasingly difficult. The factors making for moralistic extremism, and the need for compromise politics have not declined in two centuries of American independence. The last decade has demonstrated that. It clearly has been difficult to govern America in the past. It is not likely to be easier in the Third Century.

8

Public opinion
versus
popular opinion

ROBERT NISBET

OF all the heresies afloat in modern democracy, none is greater, more steeped in intellectual confusion, and potentially more destructive of proper governmental function than that which declares the legitimacy of government to be directly proportional to its roots in public opinion—or, more accurately, in what the daily polls and surveys assure us is public opinion. It is this heresy that accounts for the constantly augmenting propaganda that issues forth from all government agencies today—the inevitable effort to shape the very opinion that is being so assiduously courted—and for the frequent craven abdication of the responsibilities of office in the face of some real or imagined expression of opinion by the electorate.

Even worse is the manifest decline in confidence in elected government in the Western democracies, at all levels, and with this decline the erosion of governmental authority in areas where it is indispensable: foreign policy, the military, fiscal stability, and the preservation of law and order. For, as a moment's thought tells us, it is impossible for any government—consisting, after all, of those supposed to lead—to command respect and allegiance very long if it degrades its representative function through incessant inquiry into, and virtual abdication before, what is solemnly declared to be "the will of the people." But what is thought or cynically announced to be the will of the people so often turns out to be no more than the opinion of special-interest advocates skilled in the techniques of contrived populism—a point I shall return to later.

The important point is that from the time representative government made its historic appearance in the 18th century, its success and possibility of survival have been seen by its principal philosophers and statesmen to depend upon a sharp distinction between representative government proper and the kind of government that becomes obedient to eruptions of popular opinion, real or false. This was of course the subject of one of Edmund Burke's greatest documents, his *Letter to the Sheriffs of Bristol,* in which he declared that those who govern, once elected, are responsible only to their own judgments, not those of the electors. Across the Atlantic an almost identical position was taken by the authors of *The Federalist* and by others arguing for acceptance of the Constitution. And in a long tradition down to the present, such minds as John Adams, John Randolph of Roanoke, Calhoun, Lincoln, Tocqueville, John Stuart Mill, Sir Henry Maine, and in our own century in this country, John Dewey, Brandeis, Cardozo, and Walter Lippmann have argued along the same line.

That a just government should rest upon the consent of the governed assuredly is as true today as it was when the Declaration of Independence was signed. Equally true is the principle that the people, when properly consulted, remain the most trustworthy source of that underlying and continuing wisdom needed when great choices have to be made—above all, choices of those representatives capable of providing leadership in political matters. But to move from these truths to the position that is now becoming so widely accepted, that opinion—of the kind that can be instantly ascertained by any poll or survey—must somehow govern, must therefore be incessantly studied, courted, flattered, and drawn upon in lieu of the judgment which true leadership alone is qualified to make in the operating details of government—this is the great heresy, and also the "fatal malady" (as Walter Lippmann called it) of modern democracies.

It is worse than heresy. It is fatuous. For always present is the assumption—nowhere propagated more assiduously than by the media which thrive on it—that there really *is* a genuine public opinion at any given moment on whatever issue may be ascendant on the national or the international scene, and that, beyond this, we know exactly how to discover this opinion. But in truth there isn't, and we don't. What the eminently wise Henry Maine wrote in *Popular Government* in 1885 still seems to be true: "*Vox Populi* may be *Vox Dei*, but very little attention shows that there never has been any agreement as to what *Vox* means or as to what *Populus* means." To these words Maine added: "The devotee of democracy is in much the same position with the Greeks and their oracles. All agreed that the voice of an oracle was the voice of a god; but everybody allowed that when he spoke he was not as intelligible as might be desired."

I do not question the fact that there is in fact public opinion and that, in the modern age at least, free, democratic government must be anchored in public opinion. There is, though, as a little reflection tells us, a substantial and crucial difference between *public* opinion, properly so called, and what, following ample precedent, I shall call *popular* opinion. The difference between the two types of opinion is directly related to the differences between the collective bodies involved. Fundamentally, this is the difference between organized community on the one hand and the mass or crowd on the other.

Communities and transitory majorities

A true public, as A. Lawrence Lowell stressed in his classic work on public opinion more than a half-century ago, is at bottom a community: built, like all forms of community, around certain ends held in common and also around acceptance of the means proper to achievement of these ends. Not the people in their numerical total, not a majority, nor any minority as such represents public opinion if the individuals involved do not form some kind of community, by virtue of possessing common ends, purposes, and rules of procedure. Public opinion is given its character by genuine consensus, by unifying tradition, and by what Edmund Burke called "constitutional spirit."

Popular opinion is by contrast shallow of root, a creature of the mere aggregate or crowd, rooted in fashion or fad and subject to caprice and whim, easily if tenuously formed around a single issue or personage, and lacking the kind of cement that time, tradition, and convention alone can provide. Popular opinion is an emanation of what is scarcely more than the crowd or mass, of a sandheap given quick and passing shape by whatever winds may be blowing through the marketplace at any given time. It would be incorrect to say that popular and public opinion are totally unconnected.

What proves to be public opinion in a community is commonly generated by popular opinion, whether in majority or minority form; but it is only through a process of adaptation or assimilation—by the habits, values, conventions, and codes which form the fabric of the political community—that popular opinion ever becomes what we are entitled to call public opinion, the opinion that is in fact more than opinion, that is at bottom a very reflection of national character.

The distinction I am making may seem abstract to some, but it is a very real distinction and it has been so regarded by a long line of observers and students of government beginning in this country with the Founding Fathers, most profoundly with the authors of *The Federalist*. Few things seem to have mattered more to the architects of the American political community than that government should rest upon public opinion, upon public consent and affirmation. But in reading the key writings of that age, we are struck repeatedly by the seriousness of the thought that was given to the true nature of the public and the means proper to the eliciting from this public the will that would be most faithful to the actual character of the public, the character manifest in the people conceived as community—or rather as a community of smaller communities—rather than as mere mass or multitude brought into precarious and short-lived existence by some galvanizing issue or personality.

Hence the strong emphasis in the Constitution and in *The Federalist* upon the whole set of means whereby government, without being in any way severed from the will of the people, would respond to this will only as it had become refined through subjection to constitutional processes. Behind the pervasive emphasis in the Constitution upon principles of check and balance, division of power, and intermediate levels of government and administration ascending from local community through the states to the national government—principles which so many of the *philosophes* and then the Jacobins were to find unacceptable, even repugnant, in France when the Revolution burst there—lay a deep distrust of the human mind, of human nature, when it had become wrenched from the social contexts which alone can provide discipline and stability, which alone can put chains upon human appetites and make possible a liberty that does not degenerate into license.

There was, in short, no want of respect among the Founding Fathers for the wisdom of the people as the sole basis of legitimate, constitutional government. Neither, however, was there any want of recognition of the ease with which any community or society can become dissolved into, in Burke's words, "an unsocial, uncivil,

unconnected chaos," with destructive passion dominant where restraint and principle ordinarily prevail. There were few if any
illusions present in the minds of those responsible for the American
Constitution concerning any native and incorruptible goodness of
human nature or any instinctual enlightenment of the people considered abstractly. Steeped in the works of Thucydides, Aristotle,
Cicero, and other classics of ancient civilization and profoundly
respectful of the principles of society and government they were
able to find in the writings of Locke, Montesquieu, and Burke, the
Founding Fathers, and most particularly the authors of *The Federalist*, were well aware of the immense difference between the people conceived in terms of the social and moral attachments which
precede political organization—which indeed must underlie it if
either anarchy or despotism is to be avoided—and the people conceived in the romantic, metaphysical fashion of a Rousseau, for
whom all such attachments were but so many chains upon human
freedom.

From *The Federalist* through the works of such profound inter-
preters of the American political scene as Tocqueville, Bryce, and
Lowell, down to the writings in our own time of such perceptive
students of the political process as Lindsay Rogers and Walter Lippmann, there is a vivid and continuing awareness of the importance
of the difference I have just described: the difference between public and mere aggregate, between the people as organized by convention and tradition into a community and the people as but a
multitude, and between public opinion properly termed and opinion that is at best but a reflection of transitory majorities. It is this
awareness, forming one of the most luminous intellectual traditions
in American political thought, that I shall be concerned with in
what follows.

Federalist trust and distrust

The Federalist, for the most part originally written in the form of
individual letters to New York newspapers in 1787-88 by Hamilton,
Madison, and Jay, is by common assent the single work in American
political philosophy that can take its place among the very greatest
classics in the West since the Greeks. The unity and cogency of
the work as a whole are astonishing, given the nature of its composition; so are the comprehensiveness of scope, the social and psychological insights united with political vision, and the sheer elo-
quence. Primarily concerned with constitutional structure and

process, *The Federalist* is, among other things, a profound study of the relation of public opinion to republican government.

There is, I think, no better single insight into the *Federalist* view of the role of public opinion in government than that afforded by Number 49 of the papers. Here Madison addresses himself respectfully but negatively to the proposal, made by Jefferson, that "whenever any two of the three branches of government shall concur in opinion, each by the voices of two thirds of their whole number, that a convention is necessary for altering the constitution, or *correcting breaches of it*, a convention shall be called for the purpose."

Madison allows that there is great force in Jefferson's reasoning and that "a constitutional road to the decision of the people ought to be marked out and kept open, for certain great and extraordinary occasions." There are nevertheless, Madison writes, "insuperable objections" to Jefferson's proposal, and it is in the careful, restrained, but none the less powerful outlining of these that we acquire our clearest sense of the *Federalist* position concerning popular or public opinion.

In the first place, Madison writes, "every appeal to the people would carry an implication of some defect in the government" and "frequent appeals would deprive the government of that veneration which time bestows on everything, and without which perhaps the wisest and freest governments would not possess the requisite stability." What follows these words is central to Madison's argument and indeed to his entire political theory:

> If it be true that all governments rest on opinion, it is no less true that the strength of opinion in each individual, and its practical influence on his conduct, depend much on the number which he supposes to have entertained the same opinion. The reason of man, like man himself, is timid and cautious when left alone, and acquires firmness and confidence in proportion to the number with which it is associated. . . . In a nation of philosophers, this consideration ought to be disregarded. A reverence for the laws would be sufficiently inculcated by the voice of an enlightened reason. But a nation of philosophers is as little to be expected as the philosophical race of kings wished for by Plato.

There are two other objections Madison makes to Jefferson's proposal, both anchored in the same caution regarding the uses of public opinion. First is the serious danger of "disturbing the public tranquillity by interesting too strongly the public passions." Admittedly, we have had great success in the "revisions of our constitutions," all of which does "much honor to the virtue and intelligence" of the people. But it has to be remembered, Madison writes, that such constitution-making was at a time when manifest danger from the outside "repressed the passions most unfriendly to order and concord." Beyond this, he notes, there was the extraordinary confidence the people then had in their political leaders. We cannot, however, count on the future in this light: "The future situations in which we must expect to be usually placed, do not present any equivalent security against the danger which is apprehended."

But the greatest danger Madison foresees in any elevation of the popular will through frequent recourse to it on matters best left to the government is the unhealthy increase in legislative power, at the expense of executive and judiciary, that would inevitably follow habitual references to the people of matters of state. The legislators, Madison observes, have, by virtue of their number and their distribution in the country, as well as their "connections of blood, of friendship, and of acquaintance," a natural strength that neither the executive nor the judiciary can match: "We have seen that the tendency of republican governments is to an aggrandizement of the legislative at the expense of other departments. The appeals to the people, therefore, would usually be made by the executive and judiciary departments. But whether made by one side or the other, would each side enjoy equal advantages on the trial?"

Madison's answer is of course that they would not, that the legislators would, for the reasons just noted, tend always to outweigh the other two departments. But, he continues, even if on occasion this proved not to be the case—if, for example, the "executive power might be in the hands of a peculiar favorite of the people"—the upshot of any soliciting of popular opinion would undoubtedly be baneful. For, irrespective of where power might lie in the result, the matter would eventually turn upon not rational consideration but emotions and passions. "The *passions,* therefore, not the *reason,* of the public would sit in judgment."

How deeply Madison felt about this is attested by his repeating these arguments in *The Federalist* Number 50, where the subject is "periodical appeals to the people" rather than "occasional appeals," as in the preceding paper. Not even the institutionalization of such appeal, he thinks, would save the process from the kinds of consequences he has just described. Everything in the history of republican government suggests to Madison the ease with which issues become stripped of their rational substance and made into matters where prestige of opinion-leaders, factionalism among parties, and, not least, *passion* take command. He adduces the example of the Council of Censors which met in Pennsylvania in 1783 and 1784 to inquire into "whether the constitution had been violated." The results, Madison writes, were all that might have been expected: "Every unbiased observer may infer, without danger of mistake, and at the same time without meaning to reflect on either party, or any individuals of either party, that, unfortunately, *passion,* not *reason,* must have presided over their decisions."

Diversity and representative government

What shines through not merely Madison's thought but that of *The Federalist* generally is no simple, meretricious disdain for the people and its residual wisdom, no arrogation to some elite or natural aristocracy of the intelligence necessary to conduct government, but, instead, a solid conviction that *context* is vital in all situations where opinion and judgment are required. There are, as Madison and also Hamilton make plain, contexts in which reason and common sense will tend to come to the surface, but there are also contexts in which sheer emotions or, as Madison has it, passions dominate at the expense of rational thought. Everything possible, therefore, must be done to confine deliberations on government to the former contexts and to rely upon the vital principle of division of governmental power, of checks and balances, to maintain stability and freedom alike—hence the *Federalist* apprehensions concerning too easy, too frequent, and too regular submission of issues to the people.

It is impossible to catch the flavor of the political theory in *The Federalist,* and particularly its conception of the proper role of public opinion in government, without clearly understanding the view of human nature that was taken by Hamilton and his fellow authors. Here is no Rousseauan-romantic view of man born free and good, corrupted by institutions. On the contrary, what *The Federalist* offers us is a design of government for human beings who on occasion may be good, but who on occasion may also be evil, and for whom liberation from such institutions as family, local community, church, and government could only result in anarchy that must shortly lead to complete despotism. The essence of *The Federalist,* a notable scholar, Benjamin F. Wright, has written, "is that a government must be so constructed as to stand the strains that are inevitable. A government designed only for favorable circumstances would deserve to be rejected." And strains will exist, are bound to exist, so long as man remains what he is, invariably a compound of the good and the bad. One would look in vain for a spirit of pessimism or misanthropy in *The Federalist*. Its authors do not hate vices; they only recognize them. Edmund Burke, in his *Reflections on the Revolution in France*, would write of the French Revolutionists: "By hating vices too much, they come to love men too little." That can scarcely be said of the authors of *The Federalist*. The aim of government, free government, is simply that of providing institutions so strong, and also in such an equilibrial relationship, that neither calculated evil nor misspent goodness flowing from human nature could easily weaken or destroy them.

Where Rousseau, like so many of his impassioned fellow intellectuals in the salons of Paris, saw extermination of "factions" as the objective of government—and the extermination too, if possible, of all the smaller patriotisms in the social order in the interest of political legitimacy—*The Federalist* recognizes the inevitability of such factions and associations, with Madison declaring that "the latent causes of faction are . . . sown into the nature of man; and we see them everywhere brought into different degrees of activity, according to the different circumstances of civil society." There is to be expected a "landed interest," a "manufacturing interest," a "moneyed interest," and the like. Creditors and debtors, with their inevitably divergent interests, will always be with us. What Madison writes is: "The regulation of these various and interfering interests forms the principal task of modern legislation, and involves the spirit of party and faction in the necessary and ordinary operations of the government."

It is this recognition of the intrinsic and ineradicable diversity of the social and economic orders, of the pluralism of society, that leads the authors of *The Federalist* to their striking emphasis on *representative* institutions. Direct democracy is as foreign to the spirit of *The Federalist* as it is to the Constitution. In this, as in other respects, the philosophy of *The Federalist*'s recommendations is utterly foreign to that philosophy of government inscribed in Rousseau's political writings and in the writings of most of the *philosophes*, which lay behind the greater part of the legislation of the French Revolution. For the French radicals (and the same is also true of Bentham and the English radicals), any thought of representation was repugnant. Representative institutions were (correctly) described by Rousseau, and castigated accordingly, as "feudal" in origin. It was, happily, Montesquieu, with his virtual reverence for "mixed" government, intermediate layers of authority, and representative bodies, who proved to be the greater influence upon the Americans.

The intermediation of political authority

Montesquieu lies behind Madison's praise of political institutions as protecting society against the "diseases most incident to republican government," and behind what Madison calls "the delegation of government" to the small number of citizens elected by the rest. For pure democracy Madison has nothing but distrust: "From this view of the subject it may be concluded that a pure democracy, by which I mean a society consisting of a small number of citizens, who assemble and administer the government in person, can admit of no cure for the mischiefs of faction." Such democracies "have ever been spectacles of turbulence and contention; have ever been found incompatible with personal security or the rights of property; and have in general been as short in their lives as they have been violent in their deaths."

It is the *intermediation* of political authority, a principle that was the heart of medieval jurisprudence, as von Gierke and Maitland have told us, and that Montesquieu revived in his classic of 1748, that the authors of *The Federalist* see as vital to the proper relation of government and public opinion. Through the several, ascending layers of government, local, regional, and national, it is possible "to refine and enlarge the public views, by passing them through the medium of a chosen body of citizens, whose wisdom may best discern the true interest of their country."

Accompanying *Federalist* distrust of pure democracy, and of majorities as such, is a distrust of equalitarianism. Human beings are no more equal in their opinions on governmental matters than they are in their strengths and talents: This is the evident view of the authors, particularly of Hamilton and Madison. In Number 10, written by Madison, we find explicit statement of this: "Theoretic politicians, who have patronized this species of government [i.e., equalitarian democracy], have erroneously supposed that by reducing mankind to a perfect equality in their political rights, they would, at the same time, be perfectly equalized and assimilated in their possessions, their opinions, and their passions."

Such an argument does not repudiate the Declaration of Independence. Not even Jefferson, in whom affection was probably greatest for the doctrine of natural rights, thought that any rigid deduction could be made from the phrase, "all men are created equal." Certainly Jefferson's notable plan for public education in Virginia reveals no hint of a dogmatic equalitarianism. That human beings are equal, in moral worth at least, and that they deserve equality before the law—this was no more objectionable to a

Hamilton or a Madison than to a Jefferson. The respect in which Jefferson is held throughout *The Federalist* suggests that Madison and his fellow authors never thought that their rejection of equality as social and economic dogma repudiated the spirit of the Declaration or the ideas of Jefferson—though it would be absurd to pretend that differences did not exist between Jefferson and the authors of *The Federalist*.

There is striking similarity between the fundamental ideas of *The Federalist* and those of Burke's *Reflections on the Revolution in France,* the latter published shortly after the *Federalist* letters had made their appearance in America. When Burke wrote that "those who attempt to level, never equalize" and that "the levellers . . . only change and pervert the natural order of things," he was but echoing the sentiments contained in the passage from Madison quoted above. But it is not only concerning equalitarianism that there is substantial agreement: There is common distrust of what Burke called "governmental simplicity," common recognition of the intricate nature of man and the complexity of society, common respect for public opinion but only when duly mediated by time and institution, common veneration for representative institutions and division of authority, and common apprehension concerning mere numerical majorities, so prone, as both Madison and Burke knew, to the rise of despotism.

Tocqueville and the tyranny of the majority

Similarly, there is affinity between *The Federalist* and Tocqueville's *Democracy in America,* especially the first part, which is concerned with political institutions. We know that Tocqueville admired *The Federalist,* particularly Madison's contributions. He read the book while on his nine-month visit to this country and later with studious concentration after he had returned to Paris, and most of his interviews in this country seem to have been with Americans of definitely Federalist persuasion. Tocqueville's book is in a great many respects the child of a union effected between his French-derived interest in the democratic revolution of the early 19th century and his American-derived respect for the kind of political structures and processes which *The Federalist* had advocated in its defense of the American Constitution. For Tocqueville, as for Madison, Hamilton, and Jay, the principles of decentralization of administration, of political pluralism, regionalism, and localism, and of division of power and institutional checks upon and balances of power are fundamental to and constituent of free republican government.

So is there fundamental likeness between *The Federalist* and *Democracy in America* on the role of public opinion and on the dangers which lie in direct, popular government unmediated by the representative, deliberative bodies prescribed by the Constitution. Nearly a half-century separates the America of *The Federalist* papers from the America Tocqueville and his friend Beaumont visited in 1831. Great changes had taken place. What had been prospect for the Founding Fathers was by now reality, and as Tocqueville's *Notebooks* make clear, there was not the slightest doubt among the Americans he talked with that America's future was a secure one. It would be hard to exaggerate the buoyancy of mind, the confidence, even at times the complacency, above all the spirit of manifest progress that existed in the Age of Jackson, so far as Tocqueville's observations are concerned. And yet, hovering over all of Tocqueville's impressions and reflections on the American scene, is his concern with, his apprehensions about, the power exerted by the majority in American society, the fetters which he thought were placed upon genuine individuality by public or majority opinion.[1]

Nowhere, he writes, does public opinion rule as in the United States. There is a revealing entry, under date of October 25, 1831, in the *Notebooks*: " 'The people is always right,' that is the dogma of the republic, just as 'the king can do no wrong' is the religion of monarchic states. It is a great question to decide whether the one is more important than the other; but what is sure is that neither the one nor the other is true."

As Bryce was to point out, correctly I believe, Tocqueville exaggerated the degree of dominance by the majority in the United States that he visited, and he did not in any event ever distinguish between what he called "public opinion" and the ascendancy of the majority on a given matter. Beyond this, as political events and personages, and also literary and artistic productions within a decade or two after Tocqueville's visit, were to make incontestably clear, there was evidently not nearly the suffocating effect upon

---

[1] It will always be a matter of debate among Tocqueville scholars as to the exact proportions of the influence exerted on his mind by preoccupation with France, especially with the circumstances under which Louis Philippe had been elevated to the throne in 1830, and the influence exerted by actual experience in the United States during his short visit. As many a reviewer noted at the time, so often when Tocqueville writes "America," his eye seems to be actually on France. Nevertheless, the book, especially the first part, is about American society, and I shall treat it here in that light.

individuality in America that Tocqueville ascribes repeatedly to the majority, and (in the second part of his book) to equality. Nevertheless, Tocqueville's views on the majority are as important to us in our day as those on equality:

> When an individual or a party is wronged in the United States, to whom can he apply for redress? If to public opinion, public opinion constitutes the majority; if to the legislature, it represents the majority and implicitly obeys it; if to the executive power, it is appointed by the majority and serves as a passive tool in its hands. The public force consists of the majority under arms. . . . However iniquitous or absurd the measure of which you complain, you must submit to it as well as you can.

So impressed by and apprehensive of the majority is Tocqueville that he even compares its power to that of the Spanish Inquisition, noting that whereas the Inquisition had never been able to prevent large numbers of anti-religious books from circulating in Spain, "the empire of the majority succeeds much better in the United States, since it actually removes any wish to publish them." In the United States no one is actually punished for reading this kind of book, "but no one is induced to write them; not because all the citizens are immaculate in conduct, but because the majority of the community is decent and orderly."

Individuality and American society

It is in this context that Tocqueville utters one of his most frequently quoted observations: "I know of no country in which there is so little independence of mind and real freedom of discussion as in America." So great, he thought, was the influence of the majority's opinion upon the individual mind that the number of genuinely great or creative human beings was bound to diminish in the ages ahead. The first great generation of political leaders in the United States had, after all, been a product of different, even aristocratic, contexts. Moreover, "public opinion then served, not to tyrannize over, but to direct the exertions of individuals." Very different, Tocqueville thinks, are present circumstances. "In that immense crowd which throngs the avenues to power in the United States, I found very few men who displayed that manly candor and masculine independence of opinion which frequently distinguished the Americans of former times, and which constitutes the leading feature in distinguished characters wherever they may be found."

The effect of the majority is not merely to tyrannize the individual but also to diminish him. In the presence of the majority, Tocqueville observes, the individual "is overwhelmed by the feeling of his own insignificance and impotence." From the *Notebooks* it is evident that Tocqueville was genuinely distressed by his own observations, and by what was reported to him, of instances in which individual dissent, or, for that matter, even individual act, even though protected thoroughly by law, could be stifled by majority opinion. He refers to "the fury of the public" directed at a man in Baltimore who happened to oppose the War of 1812, and there is a long account of an interview with a white American (the gist of which went into a footnote in *Democracy in America*) on the failure of black freedmen in a Northern city to vote in a given election—the upshot of which, Tocqueville concludes, is that although the laws permit, majority opinion deprives. The majority thus claims the right of making the laws and of breaking them as well: "If ever the free institutions of America are destroyed, that event may be attributed to the omnipotence of the majority, which may at some future time urge the minorities to desperation and oblige them to have recourse to physical force. Anarchy will then be the result, but it will have been brought about by despotism."

Immediately after this passage comes a long, fully appreciative, and respectful quotation from Madison's Number 51 of *The Federalist*, in which the argument is that while it is of great importance to guard a society against the oppression of its rulers, it is equally important to "guard one part of the society against the injustice of the other part."

And yet, with all emphasis upon Tocqueville's apprehensions concerning individuality and freedom in American society as the result of majority opinion, we are also obliged to emphasize the sections of his work which deal with "the causes which mitigate the tyranny of the majority in the United States." He cites the absence of centralized administration, the presence of the frontier which made it possible for individuals to escape the conformities pressed upon them, the still-vigorous regionalism and localism of American society, the checks which executive and judiciary exert upon the majority-dominated Congress, the ascendancy of the legal profession,[2] the institution of trial by jury, and, in many ways

---

[2] " ... Some of the tastes and the habits of the aristocracy may be ... discovered in the characters of lawyers" and "in all free governments, of whatever form they may be, members of the legal profession will be found in the front ranks of all parties. The same remark is also applicable to the aristocracy."

most important for Tocqueville, the unlimited freedom of association. The latter, both in its political form of party and its civil form of interest-group, can be counted upon, Tocqueville thinks, so long as the principle remains vital, to protect individuals from the majority and the kind of government majorities seek to create.

Tocqueville was not, in sum, wholly pessimistic about the United States. He refers to it in the final pages of the first volume as the freest and most prosperous people on earth, destined to achieve a commercial and military supremacy in the distant future that will be rivaled, he writes, only by Russia. How deeply this admiration for American political institutions lay in Tocqueville's mind is evidenced by the fact that in 1848—a tormenting year for him in French politics—when he wrote a new preface to the 12th edition of *Democracy in America,* he repeated and even added to the laudatory remarks earlier written in the final sections of the first volume. If Tocqueville himself was little read after about 1880 (the present Tocqueville revival did not begin until the late 1930's), the reason is in some part that the greatness he himself had been among the first to see in American political institutions was the subject of so many works written by Americans themselves that Tocqueville's own book tended to become lost, to seem dated, even inadequate.

Bryce and the dominance of the majority

There was another classic on American government that made its appearance in 1888, at the very height of American belief in American greatness. It is doubtful that Lord Bryce's *The American Commonwealth* will ever undergo the revival Tocqueville has known in our own age, but this is in considerable degree the consequence of the very virtues of Bryce's study. It is fair to say, taking both writers in their roles as analysts of the American political scene, that so far as our own attention to Bryce is concerned, he suffers from his virtues, just as Tocqueville prospers from his defects. Bryce's command of the details of American government and its surrounding society is far superior to Tocqueville's, but it is this very command that today makes him seem dated. Tocqueville's defects of observation are only too well known, but they are offset by abstract reflections, often at the level of genius, on democracy *sub specie aeternitatis.* Fewer such reflections are to be found in Bryce, and Woodrow Wilson, then still a professor, was right in declaring them inferior to Tocqueville's. All the same, there is a significant amount of political theory or philosophy in Bryce, not least on the subject of public opinion.

Bryce's America is of course a vastly different one from that Tocqueville and Beaumont had travelled through 60 years before. The Civil War, the eruption of corporate capitalism, the unbroken expansion to the Pacific, the mushrooming of towns and cities, the development of the two large political parties and their machines, the whole public education movement, the establishment of colleges and universities everywhere in the country, the explosion of newspapers (some of great power), the spirit of nationalism, and, not least, the rise of America as a recognized world power—all of this made any treatment of American democracy in the 1880's necessarily different from anything possible in the Age of Jackson.

There are nevertheless interesting continuities to be seen in Bryce which reach back through Tocqueville to Hamilton and the other authors of The Federalist, a work that Bryce clearly admires as greatly as Tocqueville did. For Hamilton, Bryce reserved words of sheer eulogy, giving him status along with Burke, Fox, Pitt, Stein, von Humboldt, Napoleon, and Talleyrand as one among the greatest group of statesmen any single period of Western history has ever come up with. But his appreciation of the other authors of The Federalist is scarcely less, and the same holds for the principles which brought that volume into being.

What Bryce adds to The Federalist and to Tocqueville on the subject of public opinion is, first, a degree of systematic analysis in strictly scholarly style that public opinion as a concept had not had before, and second, a number of distinctions which we would not expect to find in the earlier works and which, so far as I can see, laid the essential ground on which all subsequent studies of American public opinion have been made. The section, 12 chapters in length, that Bryce gives us would have been worthy even then of separate publication, and I frankly don't think we have reached the point even yet in our knowledge of the subject that would make the reading of Bryce superfluous.

"In no country is public opinion so powerful as in the United States; in no country can it be so well studied." These opening words are followed by an inquiry into the nature of public opinion, its relation to government in earlier times and other countries, and then by a series of chapters on opinion, majority, processes of diffusion, and controlling influences upon opinion as these are to be found in the United States. In words which distinctly resemble those Walter Lippmann wrote a generation later on "stereotypes,"

Bryce tells us: "Everyone is of course predisposed to see things in some one particular light by his previous education, habits of mind, accepted dogmas, religious or social affinities, notions of his own personal interest. No event, no speech or article, ever falls upon a perfectly virgin soil: The reader or listener is always more or less biased already." Orthodox political theory, Bryce observes, assumes that every citizen has or ought to have "thought out for himself certain opinions. . . . But one need only try the experiment of talking to that representative of public opinion whom the Americans call the 'man in the cars' to realize how uniform opinion is among all classes of the people."

Yet Bryce, unlike Tocqueville, finds no tyranny of a majority. We may, he suggests, look for evidences of this tyranny in three places: Congress, the statutes of the states, and in the sentiments and actions of public opinion outside the law. But in none, he concludes, is there in fact manifestation of the dominance of the majority. Too many checks exist upon this majority in all three spheres. Bryce is skeptical indeed that such majority tyranny existed in America even in Tocqueville's day, noting dryly the great efflorescence of individuality in so many sectors of American life shortly after Tocqueville's visit.

But, Bryce continues, even if we assume that such majority tyranny did in fact exist in Tocqueville's America, a great many things have happened to check or disperse it. When Tocqueville visited the United States, the nation "was in the heyday of its youthful strength, flushed with self-confidence, intoxicated with the exuberance of its own freedom. . . . The anarchic teachings of Jefferson had borne fruit. Administration and legislation, hitherto left to the educated classes, had been seized by the rude hands of men of low social position and scanty knowledge."

Very different, Bryce writes, is the America of the 1880's. The dark and agonizing issue of slavery has been removed through the Civil War—an event, Bryce believes, that purged the American nation of many issues on which ruthless majorities were willing to ride roughshod over the rights of individuals and minorities during the years leading up to the war: "The years which have passed since the war have been years of immensely extended and popularized culture and enlightenment. Bigotry in religion and everything else has broken down." He continues: "If social persecution exists in the America of today, it is only in a few dark corners. One may travel all over the North and the West, mingling with all classes . . . without hearing of it."

In no respect, Bryce observes, is there to be found in America the kinds of violence or repression against unpopular opinions which one still finds in Ireland, France, and Great Britain. On balance, Bryce believes in the first place that Tocqueville had misperceived the nature of majority, and had greatly underestimated the authentic willingness of minorities to submerge themselves and their views in a national consensus, and in the second place that a signal change had taken place in America and, with this, in the whole structure of public opinion.

The "fatalism of the multitude"

In one respect, however, Bryce can be seen as almost a pure reflection of Tocqueville, and that is in his notable chapter on "the fatalism of the multitude." It is this fatalism—one found, Bryce argues, in all large populations characterized by "complete political and social equality"—that Tocqueville and others have confused with "tyranny of the majority." As the result of such fatalism (we would today describe it in the language of mass conformity), individuals and groups are led to acquiesce in numbers, to abandon personal and sectarian beliefs for the sheer relief of participating in the special kind of community and of power that great, undifferentiated masses represent: "This tendence to acquiescence and submission, this sense of the insignificance of individual effort, this belief that the affairs of men are swayed by large forces whose movement may be studied but cannot be turned, I have ventured to call the Fatalism of the Multitude."

Anyone who knows Tocqueville will recognize instantly that Bryce's "Fatalism of the Multitude" is indeed different from what Tocqueville had called "tyranny of the majority" in the first part of *Democracy in America*. But it is not different from, it is almost pure reflection of, the social mass—undifferentiated, monotonous, enveloping, and uniform—that Tocqueville describes in such detail in the second part of his work, the part that is devoted to equality rather than the majority. If it had been the majority in the first part that had, for Tocqueville, extinguished individuality, it is equality in the second. And in this he and Bryce are in close accord. But like Tocqueville, Bryce finds in America a large number of forces—among others, localism, regionalism, voluntary association, faith in institutions and also in the future of the country, and persisting freedom of discussion, all factors that Tocqueville had cited —which moderate this fatalism of the multitude, this inclination of mass populations to favor uniformity and the resultant sterilization of individuality.

These are the forces too which, in Bryce's judgment, give some degree of security to bona fide public opinion and its necessarily slow and deliberate spread through American society—security against the effects of suddenly formed movements and crusades. Bryce was deeply impressed by the still regnant localism and regionalism of opinion in the America of his day—the existence of, say, a profound antipathy in California toward Orientals that most of the rest of the country could only regard as bizarre, to say the least. But he was also struck by the long-run tendency of localisms and regionalisms of opinion to become fused, the elements in common to become the real stuff of American public opinion, which he did not doubt could become powerful:

> So tremendous a force would be dangerous if it moved rashly. Acting over and gathered from an enormous area, in which there exist many local differences, it needs time, often a long time, to become conscious of the preponderance of one set of tendencies over another. The elements of both local difference and of class difference must (so to speak) be well shaken up together and each part brought into contact with the rest, before the mixed liquid can produce a precipitate in the form of a practical conclusion.

Lowell and the political community

It was left to A. Lawrence Lowell, writing 20 years later at Harvard, to develop a crucial point that neither Tocqueville nor Bryce had given emphasis or focus to: the necessity of a genuine *political community,* and with this of a clearly perceived *public interest,* as the context of public opinion worthy of the name. Lowell's *Public Opinion and Popular Government,* published in 1913, is somewhat neglected these days—which is a pity, for its essential themes remain highly pertinent. The book is, so far as I am aware, the first systematic treatise on public opinion, the first to lift the subject from the ancillary if important position it has in the works of Tocqueville, Bryce, and A. V. Dicey's slightly earlier work on law and opinion in 19th-century England, and to give it a virtually prior role in the study of the governmental process. Written more or less as a textbook, Lowell's work is actually a scholarly, ground-breaking inquiry fully the equal in intellectual substance of his more famous, earlier study of English political institutions. It is this work that makes explicit the distinction I have used in this essay between public and popular, or merely majority, opinion.

It is not strange, taking the historical context into consideration, that Lowell should have made the political community a paramount consideration, for the America of his day had become increasingly torn by economic, class, regional, and ethnic interests which were being translated into political expression. Many of the voices generated by the passage of America from a predominantly agricultural to an industrial society, and by an economic system that could seem to a great many participants and observers to be in the hands of the great trusts and monopolies, were by Lowell's time becoming clamant. Populism, Progressivism, an increasingly active labor movement, socialism—at least in tractarian form, and here and there in organizational shape—and an increasingly reform-oriented Democratic Party were among the realities of Lowell's day. Woodrow Wilson, after all, had been elected on a platform of "The New Freedom" the year before Lowell's book appeared. That period of Good Feeling which Bryce had been so struck by in the period from the 1870's to the 1890's seemed suddenly to be ending, its place taken by one in which strife between parties, classes, and sections of the country over such matters as control of wealth and property, tariffs, workmen's wage and hour and safety laws, direct tax on income, and a host of others, could appear almost endemic. This was also the period in which a variety of proposals for direct, popular democracy were becoming law in many parts of the United States: the initiative, the referendum, and the recall, among other such innovations.

It is impossible to miss in Lowell's book, its scholarship and objectivity notwithstanding, an undercurrent of deep concern that *public opinion*, rooted in the people as genuine national community, generated by deeply held convictions and sentiments, and forming the necessary substratum of any free, representative government, would become confused with and blurred by mere *popular opinion*, the kind of so-called collective opinion that can be had from any person on any subject, however complex or remote, merely by the asking. The following passage from Lowell might well have been written by one of the authors of *The Federalist*:

A body of men are politically capable of public opinion only so far as they are agreed upon the ends and aims of government and upon the principles by which those ends shall be attained. They must be united, also, about the means whereby the action of the government is to be determined, in a conviction, for example, that the views of a majority—or it may be some other portion of their numbers—ought to prevail; and a political community as a whole is capable of public opinion only when this is true of the great bulk of the citizens.

Communities and legitimate majorities

Lowell is the first to define specifically the public, properly so called, as a community, one consisting—as does any form of community—of common values, ends, and acceptance of means, and endowed with the capacity to create a sense of membership and to generate belief in the reality of common or public interest. Such opinion is rooted, Lowell stresses, not simply in political motivations alone but in the lives of individuals as revealed in the full gamut of social, economic, and moral existence. Whatever may be the origin of a given expression of public opinion—in the ideology of a minority, a single party or even sect, or in even the presence of a signal personage—its reality and ultimate power as public opinion take shape only through assimilative processes whereby belief or conviction becomes bred into family, neighborhood, religion, job, and the other contexts of individual life and thought. At any given moment there may be dozens, hundreds, of popular impressions or sentiments, all capable of being voiced by one or another interest or ideology. But few of these ever become transposed into the substance of genuine public opinion. For that, time and also historical circumstance are required.

Lowell is particularly concerned with stressing the difference between a public and a mere majority. "When two highwaymen meet a belated traveller on a dark road and propose to relieve him of his watch and wallet, it would clearly be an abuse of terms to say that in the assemblage on that lonely spot there was a public opinion in favor of a redistribution of property." Lowell's example is a homely one, but it is given enlarged and pertinent significance just afterward in some words aimed at the political state: "May this not be equally true under organized government, among people that are for certain purposes a community?" In sum, a majority of voters is easily imaginable in support of redistribution of property in society or of any of a large number of proposals affecting the very social and economic base of human life. It does not follow, however, from Lowell's point of view, that such a majority necessarily reflects public opinion. For him public opinion is limited, as it was indeed for *The Federalist*, to expressions of views relating to a community and to the purposes for which the community is founded. There are legitimate and illegitimate majorities so far as the state is concerned. And a great deal that passes for "public opinion" in the judgments of interested individuals may be, and often is, no more than "popular opinion," something, as we have seen, inherently more superficial, ephemeral, and transitory.

True public opinion, Lowell adds, in words taken from his contemporary, Arthur Hadley, is composed of beliefs a man is prepared "to maintain at his own cost," not simply "at another's
cost." For any of us there is a wide range of matters on which, if
pressed, we are willing to vouchsafe a "yes," "no," or "maybe":
matters often extending into the most recondite and specialized
realms of knowledge or experience. But the only judgments which
really count, so far as genuine belief—individual or public—is concerned, are those which are, in Justice Holmes's words, "out of experience and under the spur of responsibility." These are, in the
aggregate, public opinion, and—alas, for analytical purposes—are
not commonly worn on the sleeve, not easily given verbal expression on the spur of the moment, least of all in simple affirmatives
and negatives. "Habits of the heart," Tocqueville had called them;
and they are precisely what Burke earlier had epitomized in the
word "prejudice," with what Burke called its "latent wisdom," which
engages the mind "in a steady course of wisdom and virtue, and does
not leave the man hesitating in a moment of decision, skeptical,
puzzled, and unresolved."

Lowell was clearly troubled by the problem of ascertaining public opinion in advance of its expression through regular, legitimate
political processes. There are, as he notes, individual leaders in
politics with superior gifts in this respect; they are the stuff of
which the Washingtons and Lincolns are made. But Lowell is no
more confident than either Bryce or Dicey, or Maine in his *Popular
Government* had been that there exists, or even can exist, any quick
and certain means of finding one's way to public—in contrast to
ordinary popular—opinion. Polls and surveys did not exist in Lowell's day, but I think it evident, reading him carefully at the present
time, that he would have been at least as skeptical of these as a
few later students, notably Lindsay Rogers in his classic *The Poll-
sters,* published in 1949, have been. I do not doubt that Lowell
would have approved of Rogers words: "So far as the pollers of
public opinion are concerned, the light they have been following
is a will-o'-the-wisp. They have been taking in each other's wash-
ing, and have been using statistics in terms of the Frenchman's
definition: a means of being precise about matters of which you will
remain ignorant."

Lowell knew, as had Hamilton and Madison, as had Tocqueville
and Bryce, that language as often conceals as it communicates.
Hence the untrustworthiness, or at least the precariousness, of verbalized responses to verbalized questions concerning matters of

profoundest moral, social, and political significance. We have all been struck by the shifting character of response to persisting issues as revealed in polls. But without trying to consecrate *public* opinion, I think it fair to say that this shifting, kaleidoscopic character is in fact one aspect of *popular* opinion, as mercurial in nature as the fashions, fads, and foibles which compose it. By the very virtue of its superficiality, its topical and *ad hoc* character, popular opinion lends itself to facile expression, in the polls as well as in drawing rooms and taverns, and hence, as is the case with all fashions, to quick and often contradictory change. Very different is public opinion: It changes, to be sure, as the history of the great moral and political issues in America and other nations makes evident. But change in public opinion tends to be slow, often agonizing, and—in the deepest realms of conviction—rare. The greatest political leaders in history have known this; hence their success in enterprises which, on the basis of soundings of merely popular opinion, might have seemed suicidal.

"Public interest populism"

It is the ease with which popular opinion can be confused with public opinion that accounts in substantial degree for not only the polls in American public life but also the great power of the media. The impact, the frequently determining influence of television commentators, newspaper editors, reporters, and columnists upon individual opinions is not to be doubted. In the scores of topics and issues dealt with by the media daily, the shaping, or at least conditioning, effect of the media is apparent, certainly so far as popular opinion is concerned. In this fact lies, however, a consequence that would not, a couple of decades ago, have been anticipated by very many makers of opinion in America: the rising disaffection with, even hatred of, the media in public quarters where, though the matter may not be given verbal articulation, it is believed that the media are flouting, not reflecting, *public* opinion. There should be for a long time an instructive lesson in the overnight conversion of Spiro Agnew (before the fall) from nonentity to near-hero as the result of sudden and repeated attacks upon the media, particularly television. How do we explain the fact that a medium to which tens of millions of people are drawn magnetically night after night, one that manifestly has a conditioning effect upon national thought and behavior, should face wells of potential hostility in the public, a hostility only too easily drawn upon by the right kind of political presence? Only, I suggest, through distinction between popular opinion and public opinion, difficult as this distinction may be to identify in concrete cases.

There is also what Irving Kristol has admirably described as "public interest populism," a phenomenon also, I suggest, to be accounted for in terms of popular opinion. Such populism can, as we have learned, be utterly at odds with the sentiments of large majorities, and yet, through the always available channels of popular opinion—newspapers and television, preeminently—take on striking force in the shaping of public policy. In only the remotest and most tenuous fashion, if indeed at all, does the uproar about the C.I.A., or Congress' acceptance of the H.E.W. ban on single-sex physical education classes accord with the fundamental values of American public opinion. Given, however, the variety and ingenuity of means whereby a popular opinion can be created overnight, given credence by editorial writers, columnists, and television commentators, and acquire the position of a kind of superstructure over genuine public opinion, it has not proved very difficult for a point of view to assume a degree of political strength that scarcely would have been possible before the advent of the media in their present enormous power.

A great deal of the recent turning of literally thousands of college students to law schools can be explained precisely in terms of this "public interest populism." The Warren Court first, then the judiciary as a whole, have proved to be often fertile contexts for the achievement of ends, some of them revolutionary in implication, which almost certainly would not have been achieved had they been obliged to wait for changes in American public opinion expressed through constitutional legislative and executive bodies. Mandated busing for ethnic quotas will serve for a long time as the archetype of this peculiar kind of populism. The Founding Fathers thought, and accordingly feared, legislatures in this light. We have learned, though, that the executive and the judiciary can only too easily become settings of actions which run against the grain of public opinion.

Walter Lippmann and "The Public Philosophy"

It is useful to conclude this essay by reference to a work that deserved better than its fate in the hands of most of its reviewers when it was published in 1955: Lippmann's *The Public Philosophy*. The author's interest in the subject of public opinion was doubtless generated while a student in Lowell's Harvard. His epochal *Public Opinion*, which came out in 1922, has many points of similarity with Lowell's own views and also those of Bryce. To this day, Lippmann's *Public Opinion* remains the best known and most widely read single book on the subject. Whatever may be its roots in earlier thinking, it possesses a striking originality and cogency. Even with all that has mushroomed since the 1930's in the field of the study of public opinion, Lippmann's work of half a century ago continues to offer a valuable insight into the nature and sources of public opinion in the democracies.

Even so, I prefer to deal here with Lippmann's later work, *The Public Philosophy*, for it is here, to a degree not present in the more analytical and discursive *Public Opinion*, that we find this eminent journalist-philosopher reviving and giving pertinence to the tradition that began with *The Federalist*, and that has such exemplary statements in the other works I have mentioned. I would not argue that the book is without flaw, chiefly in the difficult enterprise of trying to describe precisely and concretely the public and its genuine manifestations. But such flaws apart, it is a profound and also courageous restatement of a point of view regarding representative government that began with such minds as Burke and Madison.

"The people," Lippmann writes in words reminiscent of the apprehensions of *The Federalist*, "have acquired power which they are incapable of exercising, and the governments they elect have lost powers which they must recover if they are to govern." What, we ask, are the legitimate boundaries of the people's power? Again it could be Hamilton or Madison rather than Lippmann responding: "The answer cannot be simple. But for a rough beginning let us say that the people are able to give, and to withhold, their consent to being governed—their consent to what the government asks of them, proposes to them, and has done in the conduct of their affairs. They can elect the government. They can remove it. They can approve or disapprove of its performance. But they cannot administer the government. They cannot themselves perform."

Lippmann draws a distinction respecting the public, or people, that has been present in Western thought since the very beginning of the tradition I have been concerned with. It is the distinction between the people as mere multitude or mass, a sandheap of electoral particles, and, to use Lippmann's phrasing, "*The People* as a historic community":

> It is often assumed, but without warrant, that the opinions of The People as voters can be treated as the expression of *The People* as

a historic community. The crucial problem of modern democracy arises from the fact that this assumption is false. The voters cannot be relied upon to represent *The People*. . . . Because of the discrepancy between The People as voters and *The People* as the corporate nation, the voters have no title to consider themselves as the proprietors of the commonwealth and to claim that their interests are identical with the public interest. A prevailing plurality of the voters are not *The People*.

It is easy enough to caricature that statement and to draw from it a variety of uses, some without doubt immoral and despotic in character. In politics as in religion and elsewhere, many a leader has at times justified arbitrary and harsh rule by recourse to something along the line of what Lippmann calls "*The People*," the people, that is, as a historic, tradition-anchored, and "corporate" nation rather than as the whole or a majority of actual, living voters. None of this is to be doubted. And yet, however difficult to phrase, however ambiguous in concrete circumstance, the distinction may be— it is, I would argue—a vital one if, on the one hand, liberty is to be made secure and, on the other hand, the just authority of government is to be made equally secure.

In truth, Lippmann's distinction is but a restatement of the core of an intellectual tradition going back at very least to Burke's famous description of political society as a contract between the dead, the living, and the unborn. That description too was capable, as Tom Paine made evident, of being pilloried and mocked, of being declared a mere verbal mask for opposition to all change or a rationalization of government policy flouting the interests of the governed. And yet it is, as is Lippmann's, a valid, even vital, distinction, one that lies at the heart of a philosophy—so often termed "conservative," though it is in fact liberal—which in the 20th century, under whatever name, we have discovered to be the only real alternative to the kinds of awful power which are contained in "people's governments" or are, in our own country, hinted at in declared programs of "common cause" populism. The distinction which Burke and Lippmann make between the two conceptions of "the people" is fundamental in a line of 19th and 20th century thought that includes Coleridge, Southey, John Stuart Mill, Maine, Tocqueville, Burckhardt, Weber, and, in our own day, Hannah Arendt, Bertrand de Jouvenel, and Jacques Ellul. Basically, it is a distinction between constituted society and the kind of aggregate that, history tells us, threatens to break through the interstices of the social bond in all times of crisis, the aggregate we call the mass or crowd, always oscillating between anarchic and military forms of despotism.

Paralleling this distinction between the two conceptions of the people is the distinction, as I have tried to show here, between public opinion and what I have called popular opinion. The one distinction is as pertinent to present reality as the other. We live, plainly, in a kind of twilight age of government, one in which the loss of confidence in political institutions is matched by the erosion of traditional authority in kinship, locality, culture, language, school, and other elements of the social fabric. The kind of mass populism, tinctured by an incessant search for the redeeming political personage, where militarism and humanitarianism become but two faces of the same coin, and where the quest for centralized power is unremitting, is very much with us. More and more it becomes difficult to determine what is genuinely public opinion, the opinion of the people organized into a constitutional political community, and what is only popular opinion, the kind that is so easily exploited by self-appointed tribunes of the people, by populist demagogues, and by all-too-many agencies of the media. The recovery of true public opinion in our age will not be easy, but along with the recovery of social and cultural authority and of the proper authority of political government in the cities and the nation, it is without question among the sovereign necessities of the rest of this century.

9

The end
of
American
exceptionalism

DANIEL BELL

Y EARS ago one could buy at the Rand McNally map store a curio called "The Histomap of History." Measuring about 12 inches wide and, when unfolded, about five feet long, it shows in bands of different colors and varying widths the concurrent rise and fall of empires and peoples over a period of 4,000 years. It begins in 2000 B.C., when the Egyptians are the dominant people, flanked by the Aegeans, Hittites, Amorites, Iranians, Indians, Huns, and Chinese. By 1000 B.C., the Aegeans have disappeared; the Egyptians have been narrowed to a thin river; the Hittites, after a long period of expansion, are on the verge of extinction; the Assyrians, who begin in 1400 B.C., have begun to dominate the flow of time, widening by 800 B.C. to the major force on the world-chart. And so on, through the varying fates of the Greeks, the Romans, the Goths, the Huns. . . .

The marvel of the "Histomap" is that one can read across time at any single period, or down time, following the flowing bands of color like some rushing streams that expand into wild lakes or oceans and then contract and even disappear off the page to be replaced by bands representing some new peoples and new empires. By 1800, England begins to dominate the page, and finally the United States and Russia emerge as the two dominant powers, with bands of almost equal width in 1967, which is the last date entered on the map.

Few historians have the taste or the capacity for this kind of comprehensive view. It requires a great deal of detailed knowledge or the sweep will be superficial; or it smacks of a pretension to universal history, of seeing mankind as one, which was the mark of the UNESCO conferences (and their sponsored world history) of the 1950's. Most historians are content with the monographic concentration on a single period, a set of problems, or the history of their own nations; the cultural sweep of Geyl, Huizinga, Bloch, or Braudel is rare, though there have been recent synoptic efforts to deal with Western capitalism as a whole in the new Marxist ambitions of Perry Anderson or Immanuel Wallerstein, neither of whom, interestingly, are historians.

The one American historian who ever made such a synoptic effort was the younger, crankier brother of Henry Adams, Brooks Adams, who had less literary power than his brother but is more interesting for our purposes precisely because he took as his tableau the entirety of world history. Sharing Henry's belief in the possibility of a "scientific history" whose metaphors were drawn from the physics of Lord Kelvin (on the degradation of energy) and Willard Gibbs (on the law of phases), Brooks Adams adopted a straightforward economic determinism: The fulcrum of history rested on the dominance of metals and the control of trade routes, and empires rose and fell with shifts in the control of these strategic factors, combined with the energy and character of peoples.

In 1902 Brooks Adams wrote *The New Empire,* one of four books about the character of social revolutions and the ways in which ruling elites came to supremacy and then lost the ability to maintain their rule.[1] The focus was less on the internal tensions within a society than on the contest *between* peoples, nations, and empires (the more usual concern of the 19th century), since for Adams (as for Michelet, Taine, Ratzenhofer, Mackinder, and others) history was seen as the interplay of race and economic geography.

The New Empire is itself a "histomap": A 23-page appendix lists the major points in history from 4000 B.C., when the Pharaoh conquered the Maghara copper mines, to 1897, when the economic supremacy of America is marked by the lead of Pittsburgh in the production of steel. Adams' detailed reconstruction of world economic

---

[1] The four are *The Law of Civilization and Decay* (1897), *America's Economic Supremacy* (1900), *The New Empire* (1902), and *The Theory of Social Revolutions* (1913). All were published by Macmillan, New York.

history, through some beautiful maps, is intended to illustrate his major theme: that "during the last decade the world has traversed one of those periodic crises which attend an alteration in the social equilibrium. The seat of energy has migrated from Europe to America."[2]

In a resonant peroration, he concludes:

> ... as the United States becomes an imperial market, she stretches out along the trade-routes which lead from foreign countries to her heart, as every empire has stretched out from the days of Sargon to our own. The West Indies drift toward us, the Republic of Mexico hardly longer has an independent life, and the city of Mexico is an American town. With the completion of the Panama Canal all Central America will become a part of our system. We have expanded into Asia, we have attracted the fragments of the Spanish dominions, and reaching out into China we have checked the advance of Russia and Germany, in territory which, until yesterday, had been supposed to be beyond our sphere. We are penetrating into Europe, and Great Britain especially is gradually assuming the position of a dependency, which must rely on us as the base from which she draws her food in peace and without which she could not stand in war.
>
> Supposing the movement of the next 50 years only to equal that of the last, instead of undergoing a prodigious acceleration, the United States will outweigh any single empire, if not all empires combined. The whole world will pay her tribute. Commerce will flow to her from both east and west, and the order which has existed from the dawn of time will be reversed.[3]

American uniqueness

What is striking is not the force or even acuity of Brooks Adams' statements but the fact that they cap what had for several hundred years been a well-nigh universal expectation that the United States would inherit the future. In 1726, the idealist philosopher and Angli-

---

[2] It is interesting that the idea of "social equilibrium" was the guiding idea of the sociology of Vilfredo Pareto, who began as an engineer, and who saw the processes of history as a circulation of elites reestablishing equilibrium in society; and that Pareto's ideas should have attracted George Homans, the great-nephew of Henry and Brooks Adams. For a charming portrait of the two Adamses, see *Education by Uncles,* by Abigail Adams Homans (Boston, Houghton Mifflin, 1966).

[3] *The New Empire* (1902), pp. 208-209. Since minerals and trade routes are, for Adams, the fulcrum of economic power, he writes apropos of the rising costs of coal and iron in Europe: "The end seems only a question of time. England, France, Germany, Belgium and Austria, the core of Europe, are, apparently, doomed not only to buy their raw materials abroad, but to pay the cost of transport." (p. 176) With the rise of American economic dominance, Adams foresees the likely disintegration of Russia and the eventual shift of world economic power to the Pacific.

can bishop George Berkeley wrote a poem (shortly before sailing for America, where he hoped to establish a college for the conversion of Indians) whose last stanza has often been quoted, although its source is less well known:

> The Muse, disgusted at an age and clime,
>   Barren of every glorious theme,
> In distant lands now waits a better time,
>   Producing subjects worthy fame.
>
>   *   *   *
>
> Not such as Europe breeds in her decay;
>   Such as she bred when fresh and young,
> When heavenly fame did animate her clay,
>   By future poets shall be sung.
>
> Westward the course of empire takes its way;
>   The first four acts are already past,
> A fifth shall close the drama with the day;
>   Time's noblest offspring is the last.[4]

And more than a hundred years later, Hegel, in his *Philosophy of History*, remarked:

> America is therefore the land of the future, where, in the ages that lie before us, the burden of the World's History shall reveal itself—perhaps in a contest between North and South America. It is a land of desire for all those who are weary of the historical lumber room of old Europe.

But there was also the thought that America was not just one more empire in the long chain of men's pursuit of domination, but a transforming presence whose emergence at the center of history had been made possible not only by the providential wealth of a virgin continent, but by the first successful application of a new principle in human affairs. Again, the theme was first expressed by Brooks Adams:

> American supremacy has been made possible only through applied science. The labors of successive generations of scientific men have established a control over nature which has enabled the United States to construct a new industrial mechanism, with processes surpassingly perfect. Nothing has ever equaled in economy and energy the administration of the great American corporations. These are the offspring of scientific thought. On the other hand, wherever scientific criticism and scientific methods have not penetrated, the old processes prevail, and these show signs of decrepitude. The national government may be taken as an illustration.

---

[4] The poem is reprinted in the reader, *Manifest Destiny and the Imperialism Question*, edited by Charles L. Sanford (New York, John Wiley), pp. 21-22.

And although a pedantic social scientist in the Great Exhibition Hall of History might seek to establish a morphology of societies by forms and types, the belief arose that the features of the United States were historically distinct and unrepeatable. This is the argument of Daniel Boorstin's celebrated book *The Genius of American Politics,* in which he writes:

> The genius of American democracy comes not from any special virtue of the American people but from the unprecedented opportunities of this continent and from a peculiar and unrepeatable combination of historical circumstances. . . . I argue, in a word, that American democracy is unique. It possesses a "genius" all its own.

It was an expansion of Tocqueville's theme of American uniqueness, the sense, as Richard Hofstadter has put it, "of the ineluctable singularity of American development . . . the preformed character of our democratic institutions, the importance of the democratic revolution that never had to happen."[5]

All of this added up to the conception of "American exceptionalism," the idea that, having been "born free," America would, in the trials of history, get off "scot free." Having a common political faith from the start, it would escape the ideological vicissitudes and divisive passions of the European polity, and, being entirely a middle-class society, without aristocracy or *bohème,* it would not become "decadent," as had every other society in history. As a liberal society providing individual opportunity, safeguarding liberties, and expanding the standard of living, it would escape the disaffection of the intelligentsia, the resentment of the poor, the frustrations of the young—which, historically, had been the signs of disintegration, if not the beginning of revolution, in other societies. In this view, too, the United States, in becoming a world power, a paramount power, a hegemonic power, would, because it was democratic, be different in the exercise of that power than previous world empires.

Today, the belief in American exceptionalism has vanished with the end of empire, the weakening of power, the loss of faith in the nation's future. There are clear signs that America is being displaced as the paramount country, or that there will be the breakup, in the next few decades, of any single-power hegemony in the world. Internal tensions have multiplied and there are deep structural crises, political and cultural, that may prove more intractable to solution than the domestic economic problems.

---

[5] Richard Hofstadter, *The Progressive Historians* (New York, Alfred A. Knopf, 1968), p. 445.

What happened to the American dream? Are we now caught up in the *ricorsi* of history, so that in the "histomap" of the 21st century the span of American color will have thinned to the narrow stream of a vanquished nation, yet another illustration of the trajectory of human illusions? Simply to recollect all those minds who believed, often with enormous confidence, that they had the "master key" to the course of history should give pause to anyone today intent on making incautious generalizations. What I would rather do here is retrace the course of the American belief in exceptionalism and see where we stand as we approach the third American century and the second Western millennium.

II. Manifest Destiny

A nation or a people is shaped by nature, religion, and history. Mountains or plains or seas influence the varieties of national character. Religion provides an anchorage, even when people are uprooted. History, bound by the principle of inheritance, provides a sense of distinction and of continuity, so that, as Burke put it, a society is a partnership of the living, the dead, and the unborn. In the history of different peoples it has usually been one or another of these fundaments that was predominant in shaping the distinctive character of the race.

In the United States, nature and religion intertwined to form the character of the nation. There was the awesome expanse of the land with its extraordinary variety and fertility. Equally, at the start, there was a covenant—explicit with the Puritans, implicit in the deism of Jefferson—through which God's providential design would be unfolded on this continent. There was no history but an act of will, and by that act a new people was created.

A people, as Herder defined it, is held together by the interwoven skein of language and culture in which the past is ennobled, through myth and story, to become history. In the early part of the 19th century, that extraordinary reactionary Joseph de Maistre predicted the failure of the United States because the country had no proper name, and therefore no collective identity. Yet as Orestes Brownson wrote in *The American Republic,* "The proper name of the country is America: that of the people is Americans. Speak of Americans simply, and nobody understands you to mean the people of Canada, Mexico, Brazil, Chile, Paraguay, but everybody understands you to mean the people of the United States. The fact is significant and foretells for the people of the United States a continental destiny...."

When the United States of America was proclaimed, the larger portions of the continent were held by France, Spain, and England, not by the new nation. (In 1789, Talleyrand referred to the Alleghenies as "the limits which nature seems to have traced" for the Americans.) But from the start there was a doctrine of geographical predestination, defined by either the needs of security, or political necessity, or by the contours of nature itself. That argument lay behind the Louisiana Purchase and the acquisition of Florida, which, as one writer remarked, "physiographically belonged to the United States," and, later, the annexation of Texas. It was concerning the latter that the most pregnant phrase for justifying the course of expansion was coined. The annexation of Texas, wrote John L. O'Sullivan in the *Democratic Review* in 1845, was "the fulfillment of our manifest destiny to overspread the continent allotted by Providence for the free development of our yearly multiplying millions."

Manifest Destiny was the civil religion of 19th-century America: not just the idea that a nation had the right to define its own fate, but the conviction of a special virtue of the American people different from anything known in Europe or even, hitherto, in the history of the world. The theme was first announced by Thomas Paine in *Common Sense,* in which he justified the American rebellion on the ground of a special American metaphysical destiny and mission. It received the endorsement of Ralph Waldo Emerson, who wrote: "[America] is the country of the Future . . . it is a country of beginnings, of projects, of designs, of expectations. Gentlemen, there is a sublime and friendly destiny by which the human race is guided." And it had its heraldic bard in Walt Whitman, who in millennial fashion saw America leading the human race to a new greatness. For this reason, Whitman claimed Mexican lands by "a law superior to parchments and dry diplomatic rules," the law of beneficent territorial utilization. (And he added, "Yes, Mexico must be thoroughly chastised.") In 1846, he demanded the retention of California on the ground that America's territorial increase meant "the increase of human happiness and liberty," and he further declared that while "it is impossible to say what the future will bring forth . . . 'manifest destiny' certainly points to the speedy annexation of Cuba by the United States." And in his poem "Passage to India," Whitman reached out to a vision of a superior civilization encircling the globe from East to West under the auspices of America: Celebrating the completion of the transoceanic cable, Whitman envisaged the movement of civilization from its birth to its culmination in the

West, crossing the Pacific to forge in a great circle of time a link with the ancient civilization of Asia.[6]

American imperialism

At what point Manifest Destiny merged with imperialism is a question that historians still dispute. One can cite the famous 1846 speech of Senator Thomas Hart Benton on the Oregon question: "Futurity will develop an immense, and various, commerce on that sea [the Pacific], of which the far greater part will be American. . . . It would seem that the White race alone received the divine command, to subdue and replenish the earth! For it is the only race that has obeyed it—the only one that hunts out new and distant lands. . . ." Or there was Alfred Thayer Mahan, who in 1890 advanced the thesis that the United States must enter into vigorous competition with other powers over foreign markets, build a huge navy, acquire naval bases in distant waters, and expand by acquiring colonies beyond the confines of the Western hemisphere.

Yet Benton, like Whitman, justified his doctrine on the ground of progress, and declared that the Caucasians and the Mongolians "must talk together, and trade together, and marry together. Commerce is a great civiliser—social intercourse as great—and marriage greater." And Mahan scorned the "mere dollar-and-cent view, the mere appeal to comfort and well-being as distinct from righteousness and foresight." He was speaking for his own class, men who followed a "nonpecuniary profession," and who wanted the expansion of seapower not for money-grubbing reasons, which he despised, but for national grandeur.

In 1898-99, the United States suddenly became a colonial power. It annexed the Hawaiian Islands. Defeating Spain, it took Puerto Rico and the Philippines. It acquired Guam and part of Samoa, and had there not been a delay in the Danish Rigsdag, would have bought the Virgin Islands as well. In an 18-month period it had become the master of empires in the Caribbean and the Pacific.

Although the actions were initially hailed in the religious press (as America's Christian duty), and applauded by the business world, within a year there was an almost complete turnaround in American opinion and political action. Cuba was given its indepen-

---

[6] The quotations from Whitman and many of the references in this section are taken from Albert K. Weinberg's *Manifest Destiny* (Baltimore, The Johns Hopkins Press, 1935), the magisterial study of this idea in American life and thought.

dence. Pro-annexation overtures by Haiti and Santo Domingo were rebuffed. And moves to acquire leaseholds in China were dropped. As Ernest R. May, who has chronicled this period in *American Imperialism: A Speculative Essay*, remarked: "After 1900 scarcely a Congressman or newspaper editor raised his voice in favor of further colonial extension. Imperialism as a current in American public opinion appeared to be dead."

It would be difficult to adduce a singular reason for this startling reversal of opinion, yet as May mobilizes the evidence, the basic fact seems clear: The nation's "establishment," which in 1898, as in the decade before, had been divided on the issue of imperialism, by the turn of the century had come to a new, unified point of view. As May concludes: "The American establishment once again possessed an anticolonial consensus as firm as that which had existed in the early 1880's."

The American establishment at the time consisted of the leaders of the legal profession, the universities, and the editors and publicists, centered largely in Boston and New York, who molded opinion for the country. Their stand was typified by the fact that Charles W. Eliot, the president of Harvard, was an active member of the Anti-Imperialist League, a group financed by Andrew Carnegie. The antipathy to imperialism was fed by many sources. The word "imperialism" had come into the English language in the 1860's to describe the policies of Louis Napoleon, a despot of great demagogic popularity with the masses, whose foreign adventures, such as his expedition to Indochina and the effort to place Maximilian on the throne of Mexico, were condemned by liberals everywhere. Liberalism, indeed, provided the foundation for the attack on imperialism. In classic liberalism the relevant social unit was "society," not the "state," and imperialism was essentially a continuation of mercantilist policies to enhance the power of the state. Both mercantilism and imperialism, by using the state to monopolize trade, interfered with "natural" economic processes. In this respect, the influence of Cobden and Bright in economic policy, and of Gladstone in politics, was decisive. Wealth would be increased for all through free trade, and through the doctrine of comparative advantage, which allowed those countries and peoples best equipped by resources and skills to manufacture more cheaply the goods needed by other parts of the world. Finally, the very origin of the United States as a colony of Great Britain, and the desire for freedom which had led the colonies to revolt against British rule, led many members of the establishment to reject American rule over other unwilling colonies. Manifest Destiny, as they supported the idea, was continental predestination, not overseas power.

The debates about American colonialism, however, left an ambiguous legacy in American political life. A few members of the establishment, such as Brooks Adams and Admiral Mahan, were straightforward imperialists, drawing their arguments from geopolitical considerations and a tough-minded theory of world-historical development. But most, being liberal, were not imperialists. Yet the very moralism and rhetoric which had sanctified the American mission, while drawing the country back from the idea of territorial acquisition, still impelled America to be a redeemer of the world. Whitman's tract, *The Eighteenth Presidency*, speaks of a redeemer nation and a Redeemer-President. And the words of Woodrow Wilson, seeking to persuade the country to join the League of Nations, are soaked in the rhetoric of redemption. "Nothing less depends on this decision, nothing less than the liberation and salvation of the world," he said. The world had accepted American soldiers "as crusaders, and their transcendent achievement has made all the world believe in America as it believes in no other nation organized in the modern world." In sum, "America had the infinite privilege of fulfilling her destiny and saving the world." [7]

This ambiguity about colonialism and imperialism would be magnified two decades later in the rhetoric and actions of Franklin D. Roosevelt, who worked actively to have the French and Dutch dismantle their colonial empires out of the profound conviction that colonialism was wrong, yet took steps calculated to establish American political leadership, if not hegemony, in the world as a whole. World War II was the fateful turning point for American society. In a striking way, the situation of the United States vis-à-vis the rest of the world resembled that of Athens after the defeat of the Persians. Athens had to choose between returning to an older, primarily agricultural form of life, and expanding as a mercantile power. The Athenians voted for expansion and committed themselves, under the leadership of Pericles, to a conscious policy of imperialism and democracy.

The United States, though isolationist after World War I, could not retreat to an insular role in 1945. The scope of America's economic reach was now worldwide. And if political power did not

---

[7] For a discussion of this aspect of American beliefs, see Ernest Lee Tuveson, *Redeemer Nation: The Idea of America's Millennial Role* (Chicago, University of Chicago Press, 1968), especially chapter VI. The quotations from Wilson are taken from Tuveson.

necessarily follow the contours of the expanding economic influence, it had a trajectory of its own—to fill the power vacuums created by the withdrawal of the British and French from Asia, to defend Europe itself against the pressures of Russian expansion.

The "American Century"

Yet it was not only sober considerations of world order or national interests that propelled the American destiny. There was—there almost had to be—the messianic language and the sense of mission that derived from the American character, and it is no accident that the attempt to define this role was made by Henry Luce, the son of a missionary and the proprietor of *Time*.

In 1942, Luce set up a department of Time, Inc., called the "Q" department (after the "Q" ships of World War I) to formulate proposals on the shape of the postwar world and to expand the ideas that Luce had expounded in a famous editorial in *Life*, in February 1941. In that essay, entitled "The American Century," Luce wrote:

> As America enters dynamically upon the world scene, we need most of all to seek and to bring forth a vision of America as a world power which is authentically American. . . . And as we come now to the great test, it may yet turn out that in all our trials and tribulations of spirit during the first part of this century we as a people have been painfully apprehending the meaning of our time . . . and there may come clear at last the vision which will guide us to the authentic creation of the 20th century—our Century.
>
> America as the dynamic center of ever-widening spheres of enterprise, America as the training center of the skillful servants of mankind, America as the Good Samaritan, really believing again that it is more blessed to give than to receive, and America as the powerhouse of the ideals of Freedom and Justice—out of these elements surely can be fashioned a vision of the 20th century. . . .
>
> It is in this spirit that all of us are called, each to his own measure of capacity, and each in the widest horizon of his vision, to create the first great American Century.

The theme of the American Century quickly became the object of skepticism and derision. In 1946, challenged by an English editor to provide a "self-confident announcement of what America stands for," Luce hesitated and mumbled something about "having been burnt, long ago, at that fire."[8] And not long after that, he picked

---

[8] See Robert T. Elson, *The World of Time Inc., Volume Two: 1941-1960* (New York, Atheneum, 1973), pp. 17-20. Luce had been rebuffed earlier. In 1942, he was invited to dinner with Churchill at Ditchley and, as he recorded in the draft of an unpublished book, ". . . . I veered to the question of 'postwar planning.'

up a small book by Reinhold Niebuhr, *Discerning the Signs of the Times,* in which that wise theologian had written: "Just as nationalistic and universalistic elements were present in the Messianic expectations of even the greatest prophets, so also each new nation mixes a certain degree of egoistic corruption with its more dangerous hope not only for a reign for peace but also for an 'American Century.'" And Luce noted: "Having absorbed Niebuhr, I now know about the pitfalls and heresies involved in the American Century. I think I am now no longer afraid to 'redefine the American Century.'" But as his chronicler notes, he never did so specifically, though the center of his thinking came to rest in the espousal of the world rule of law.

The American Century lasted scarcely 30 years. It foundered on the shoals of Vietnam. One can posit many explanations of the deepening American involvement there. Arthur Schlesinger has propounded the "quagmire" theory, whereby each step of aid sucked us further into the swamp and made it more difficult for us to extricate ourselves. There is the variant idea of the power vacuum: As the French were forced to withdraw we stepped in, lest the domino structure of client states collapse. And there is the conventional left-wing argument that Vietnam was an inevitable extension of American imperialism.[9]

Whatever the truth of the specific historical arguments, what is clear is that none of these explanations deals with the fundamental quality of national style and character which shaped the American actions—namely, the hubris, the "egoistic corruption" which expressed itself in the belief that America was now the guardian of world order and the United States as a matter of pride (tinted as always by moralism) had to take its "rightful" position as the leader of the free world. This was no less true of John F. Kennedy's Inaugural Speech than it was of Henry Luce's "triumphal purpose."

One can cast all this in a deterministic mold and say that the

---

The next thing I felt was a hearty slap on the back, and Churchill was saying: 'Never mind about all that, Luce. Just win the war—and then all will be well.'" But Luce at that time was undiscouraged, and the "Q" department at Time, Inc., renamed the Postwar Department, went ahead with its studies, subsequently published as supplements to *Fortune.*

[9] Whatever the plausibility of imperialism as a component of America's economic interests, the example of Vietnam would make the least sense as an area where vital or basic American economic interests were at stake.

For an interesting and critical examination of the necessity of economic imperialism as the major force in American foreign policy, see Barrington Moore, Jr., *Reflections on the Causes of Human Misery* (Boston, Beacon Press, 1972), pp. 116-132, where Moore examines the arguments of Baran, Sweezey, Magdoff, Kolko, and other Marxist writers.

centrality of the American world role was an inevitable consequence of the weakness of other states, or the inevitable rivalry
with the Soviet Union, or that the idea of Manifest Destiny and
mission inevitably would carry the United States into the moralistic
role of world policeman. Whatever the truth of these cases, the fact
is that these molds have now been broken. There is no longer a
Manifest Destiny or mission. We have not been immune to the corruption of power. We have not been the exception. To a surprising
extent there is now a greater range of choice available to the
American polity. Our mortality now lies before us.

III. The America within

In *The Great Christian Doctrine of Original Sin Defended* (1758),
Jonathan Edwards argued that depravity is inevitable because the
identity of consciousness makes all men one with Adam. As we now
see, History has traduced Manifest Destiny. The American Exceptionalism is the American Adam. Yet if destiny is no longer the sure
ground of American exceptionalism, what of those domestic conditions of American life—religion and nature—that have shaped the
American character and institutions? Can we escape the fate of
internal discord and disintegration that have marked every other
society in human history? What can we learn from the distinctive
ideological and institutional patterns that have, so far, shaped a
unique American society and given it distinctive continuity in 200
years of existence? Any specification of shaping patterns is bound
to be incomplete. What I single out are those aspects which allow
me to test, within the domestic order of the American polity, the
fate of American exceptionalism in these two centuries: Americanism, the land, equality, cultural diversity, space and security,
economic abundance, and the two-party system.

The Puritan covenant which defined the early New England
settlement was a metaphysical passion which drew its fuel from a
hostility to civilization, suppressing the springs of impulse, and
drawing human will directly from God rather than from man-made
institutions. Yet the very conditions of American life, the need for
self-reliance and the evidence that one could change the world by
one's own efforts, gradually eroded the otherworldly foundations
of Puritan New England, and stressed the need to find one's self,
one's achievements, one's salvation in the here and now. To make
one's faith center on *this* world, to reject theology and dogma and
the immemorial rituals of classical religions was, as Harold Laski

has pointed out, the central principle of Emerson's famous address to the Harvard Divinity School in 1838. The religion of America, whether we look to Emerson or Whitman, was *Americanism.*

"Americanism" meant that this was, as the Great Seal of the United States declared, a "new order of the ages," that here one could *make* one's self rather than simply continue the past or, if one came as an immigrant, *remake* one's self. It is striking that almost all of Marx's co-workers in the German Workers Club who came to the United States after 1848 (including the leader of the insurrectionary wing of the Socialist movement, August Willich, Marx's fiercest antagonist on "the left") abandoned socialism when they came to the United States. It was Hermann Kriege, a founder of the League of the Just, who declared that "Americanism" was a surrogate for his former socialism, and that free land and a homestead act would provide a permanent solution to any American social problem.[10]

Contrary to popular impression (largely created by a press looking for sensational stories), most immigrants were not radicals or agitators. As Marcus Lee Hansen pointed out many years ago, the overwhelming majority of immigrants were staunch supporters of the country and quickly became "conservative."

Americanism was a creed and a faith. As Leon Samson, a neglected socialist writer whose works have been resurrected by S. M. Lipset, wrote 40 years ago:

> When we examine the meaning of Americanism, we discover that Americanism is to the American not a tradition or a territory, not what France is to a Frenchman or England to an Englishman, but a doctrine—what socialism is to a socialist. . . . Every concept of socialism

---

[10] When Marx's friend and co-worker Joseph Weydemeyer sailed to New York in 1851, Engels wrote him a letter of caution:

> That you are going to America is bad, but I really don't know what other advice to give you if you can't find anything in Switzerland. . . . Your greatest handicap, however, will be the fact that the useful Germans who are worth anything are easily Americanized and abandon all hope of returning home; and then there are the special American conditions: The ease with which the surplus population is drained off to the farms, the necessarily rapid and rapidly growing prosperity of the country, which makes bourgeois conditions look like a *beau ideal* to them, and so forth.

Weydemeyer became a brigadier general in the Union Army in the American Civil War, as did August Willich. Weydemeyer remained a friend of Marx, but Willich and most of the other German socialists became Republicans and even held minor electoral posts, especially in Ohio, which had a German socialist concentration. For a discussion of this emigration, see Carl Wittke, *Refugees of Revolution: The German Forty-Eighters in America* (Philadelphia, University of Pennsylvania Press, 1952) and R. Lawrence Moore, *European Socialists and the American Promised Land*, (Oxford University Press, 1970). The letter to Weydemeyer is cited in Moore, pp. 4-5.

has its substitutive counterconcept in Americanism, and that is why the socialist argument falls so fruitlessly on the American ear.[11]

The central doctrine was the idea of individual achievement free of class origins; of individual mobility, geographical and social; of equality of opportunity, and the acceptance of the risks of failure. The central image was the idea of individual enterprise. These were possibilities drawn from the character of an open society, the world as pictured in the America of the 18th and 19th centuries.[12]

Yet today all such ideas must have a different meaning in a world where such individual enterprise is no longer possible, a world of organizations where 85 per cent of the labor force are wage and salary employees. To that extent there is always the problem of squaring a new reality with an old ideology, or of redefining or giving a different meaning to the idea of achievement (e.g., the hope of business corporations that its members will identify achievement with the *corporate* enterprise, not the individual—a corporate identity which does take place, say, in Cuba or China).

The larger question however is the absence of a faith or a creed. Do most Americans today believe in "Americanism"? Do people identify the doctrine of achievement and equality with pride in nation, or patriotism? It is an open question.

The land

In the beginning was the land.[13] It was this providential Eden "that God hath espied out . . . for Him" (as John Cotton put it)

---

[11] "Americanism as Surrogate Socialism," reprinted in *Failure of a Dream?*, edited by John H. Laslett and S. M. Lipset (New York, Doubleday, 1974), p. 426. The essay appeared originally in the book *Toward a United Front* (New York, Farrar and Rinehart, 1935).

[12] In a footnote in *Capital* Marx cites as an illustration of the varieties of work which should be available to a man—lest he be a "detailed worker, crippled by life-long repetition of one and the same trivial operation, and thus reduced to the mere fragment of a man"—the experiences of a worker in the new world:
A French workman, on his return from San-Francisco, writes as follows: "I never could have believed, that I was capable of working at the various occupations I was employed on in California. I was firmly convinced that I was fit for nothing but letterpress printing. . . . Once in the midst of this world of adventurers, who change their occupation as often as they do their shirt, egad, I did as the others. As mining did not turn out remunerative enough, I left it for the town, where in succession I became typographer, slater, plumber, &c. In consequence of thus finding out that I am fit for any sort of work I feel less of a mollusk and more of a man" (A. Courbon, *De l'enseignement professionel*, 2eme ed., p. 50). [*Capital*, vol. I, p. 534]

[13] I take this, and several other items in this inventory, from Robert Wiebe's *The Segmented Society* (Oxford University Press, 1975), though at variance with his interpretation, and with different illustrations.

that made the first settlers create the great romance of the American wilderness. As Daniel Boorstin writes:

> The magic of the land is a leitmotif throughout the eighteenth and nineteenth centuries. We hear it, for example, in Jefferson's ecstatic description of the confluence of the Potomac and Shenandoah rivers; in Lewis and Clark's account of the far west; in the vivid pages of Francis Parkman's Oregon Trail; and in a thousand other places. It is echoed in the numberless travel-books and diaries of those men and women who left comfortable and dingy metropolises of the Atlantic seaboard to explore the Rocky Mountains, the prairies, or the deserts.

But the land was also a shaping element on its own. As Frederick Jackson Turner wrote: "American democracy was born of no theorist's dream. . . . It came out of the American forest and it gained strength each time it touched a new frontier." Frontier democracy was natural. It evoked a "fierce love of freedom, the strength that came from hewing out a home, making a school and a church, and creating a higher future for his family." This conception, he said in 1903, "has vitalized all American democracy and has brought it into sharp contrasts . . . with those modern efforts of Europe to create an artificial democracy by legislation." In Turner's view, therefore, democracy in America was naturally a condition of a mental climate born of the physical environment, whereas in Europe it was an artificial contrivance imposed on the environment and not implanted there by nature. As Turner concluded from this contrast: "Other nations have been rich and powerful, but the United States has believed that it had an original contribution to make to the history of society by the production of a self-determining, self-restrained, intelligent democracy. It is in the Middle West that society has formed on lines least like Europe. It is here, if anywhere, that American democracy will make a stand against a tendency to adjust to a European type."

Like so many such visions, the "cosmology" is derived from an agrarian society. But in a world today where few people work "against" nature—on the land, in the forests, in the mines, or on the seas—where work, particularly in a post-industrial society, is largely a "game between persons," in which nature and things are excluded from daily life, what is the meaning, or shaping character, of the land? The sense of "unspoiled grandeur" still gives passion to the drive of environmentalists to stay the destruction of forests and wetlands. And the land still retains a romance for those who want to "drop out" and live (for a few years) in the comparative isolation of Vermont or Maine. But the land, by and large, is an economic spoil, cut up, with few controls, into gridiron lots for suburban development or recreation retreats. And even where the awesome vistas remain (once one can get away from thousands of cars piling into the national parks), it is now only "out there," a view to be admired, and no longer a shaping element of its own.

Equality and cultural diversity

The idea of equality in America has its roots in mythic soil. "'Since becoming a Real American,' roared Paul Bunyan, 'I can look any man straight in the eye and tell him to go to hell! If I could meet a man of my own size, I'd prove this instantly. We may find such a man and celebrate our naturalization in a Real American manner. We shall see. Yay, Babe!'" These were the sentiments of Paul and his pal as they stood before the Border, and then leaped over to become Real Americans.

They are also the observations of European travelers, applauding or appalled, as they observed the free-and-easy ways of Midwestern Americans, the unwillingness to "doff one's cap" or use the deferential "sir." It is the oldest cliché, and truth, about the American image, if not the actuality. For my colleague Samuel Huntington, the "challenge to authority" is the underlying factor of the problem of governability in democracy today. And its source is the recurrent populism, the frontier egalitarianism, which has been the demagogic appeal of American politics since the days of Andrew Jackson, and the Cider Barrel election of 1840. Yet that rough-and-ready egalitarianism has also gone hand in hand with another swaggering attitude in which the "top dog" is going to show the underling "who is boss." The idea of the "boss," whether on the job or in the political machine, has also been a staple of American life. The two ideas have not been contradictory because the emphasis remained on the individual.

Where there is a difference today, it is that authority in a technical and professional society is necessarily vested in *acquired competences* and *impersonal attributes*, not in the *personal qualities* of the individual. It is this erosion of the immediate, the personal, and the individual, and the rise of bureaucratic authority, which lead to so much irritation and disquiet. In the United States, the tension between liberty and equality, which framed the great philosophical debates in Europe, was dissolved by an individualism which encompassed both. Equality meant a personal identity, free of arbitrary class distinctions. It is the loss of that sense of individuality, promised by equality, which gives rise to a very different populist reaction today, both among the "left" and the "right," than in the past.[14]

There is, equally, a disorientation because of the breakup of cultural diversity. The differences in America were regional and religious, differences of speech, custom, and manner summed up in such stereotypes as a New England Yankee, a Virginia gentleman, a Midwest farmer, a Texas rancher, or any other of a dozen images from the Frank Capra movies, the songs of Woody Guthrie, or the maunderings of Studs Terkel.

Here again, repetition has dulled our awareness of reality. People *were* different, their differences derived from cultural heritage, generations of immigration, the character of local communities, occupational habits, religious practices, and the like. The destiny of America, Harold Laski wrote in 1948, is still in the melting pot, the creation of a homogeneous people so that Americanism would mean the same to a sharecropper in Arkansas, a steelworker in Pittsburgh, and a farmer in Kansas.

But the melting pot has yielded its meld. America today is homogeneous: not in the superficial existence of a national popular culture created by television (*Gunsmoke* and its demise do make a common conversational gambit for persons in any and all parts of America), but in the very fact of a hedonism which is the common value—in the idea of consumption and of exhibition—that unites middle-class and youth cultures alike, and which irons out the differences in life-styles and habits in the country.

The resurgence of ethnicity, which has been so marked in recent years, is not a new concern with cultural diversity (the only example of cultural "differences" are ethnic food fads which are quickly absorbed into middle-class homes) but a political strategy, a means whereby disadvantaged groups use the political process to claim a share of the goods that are created by the homogeneous hedonistic culture.

It is this very cultural homogeneity that marks a new crisis of consciousness, for we have become, for the first time, a common people in the hallmarks of culture. Even the old distinction of "highbrow" and "lowbrow," which Van Wyck Brooks installed 60 years ago and which was pursued so vigorously 20 years ago by Dwight Macdonald (who added the category of the "middlebrow"),

---

[14] I leave aside the very different question of the conflict that has arisen between the principle of "equality of opportunity" and the desire for an "equality of results," or the translation of the demands for equality into the claims of entitlement. I have discussed these questions in my essay, "The Public Household," in *The Public Interest*, No. 37 (Fall 1974).

has lost its meaning today. Are *M*A*S*H** and *Nashville* high-brow or lowbrow? In fact, neither: They are Middle America mocking itself in the accents of the highbrow and the lowbrow. Yet despite a common culture, there is no common purpose, or common faith, only bewilderment.

Space and security

The United States, unlike most major powers in the world, has enjoyed a unique freedom from both immediate military threats and the experience of invasion. Since the War of 1812, no foreign armies have fought on American soil. We have not had a large standing army or a military caste and, until World War II, no continuing draft of young men for extended service in the Army. Large geographical distances and the difficulties of long-distance logistics made space an effective factor in American security. As Robert Wiebe remarks: "Security relaxed the social fabric. Simply and profoundly, freedom from military imperatives meant freedom to go about one's affairs ... Throughout its history, in other words, America had escaped a fundamental part of life almost everywhere else around the globe."

Yet there *was* internecine conflict. Apart from the Civil War, with its deep tear in the social fabric, the history of the country has been marked by an extraordinary amount of violence—frontier battles in the West, grave labor strife that raged for almost 75 years, and crime in the cities. Yet here too, space placed invisible and real barriers between such violence and both the political life of the country and the daily lives of individuals. In the cities, crime was marked off geographically, being restricted largely to the port areas and the slums; in a curious sense, the "dangerous classes" knew their place and battered each other, leaving the segregated middle- and upper-class areas peaceful and calm. Frontier violence was pushed steadily westward, as the boundaries and marginal occupations moved across the country; and in the inevitable cycle of routinization, the small towns settled down to mundane economic life. And the remarkable fact about labor violence was that, while it was more explosive and intense (involving dynamiting, gun battles, and the use of troops and police) than in the ideology-riven countries of Europe, this violence (in the coal mines, the timber camps, the textile mills) took place largely at the "perimeters" of the country. It took a long while for these shock waves to reach the political center, and by that time their force had been dissipated.

What saved this country from internal disorder was not so much the "lack of ideology" as the insulation of space.

The contemporary revolutions in communication and transportation—television and jet airplane—have meant, geographically, an "eclipse of distance." In 1963, when A. Philip Randolph and Martin Luther King planned for a March on Washington, within 48 hours almost 250,000 persons had flowed into the capital to stand on the Mall, within sight of the President's office, to voice their demands for civil rights legislation. During the Vietnam War, "marches" of up to 70,000 demonstrators repeatedly stormed into Washington. The last such mass protest, spurred by the "Mayday Tribe," resulted in a series of actions to blockade the bridges leading into Washington from Virginia—actions that were halted only by the wholesale arrest of more than 5,000 persons, arrests which later prompted civil suits against the government and a judicial ruling that those arrested were entitled to pecuniary restitution from the government. (To that, at least, one can still say, "Only in America.")

The simple point is that God's gift of insulated space has disappeared. The United States is no longer immune to the kind of "mobilization politics" that has been characteristic of Europe in the past and of almost every other country in the world today. Mobilization politics, by its very nature, organizes direct mass pressure on a political center. What made France a political hotbed was the concentration of power in Paris, surrounded by a "Red belt" of workers in such *banlieus* as Billancourt, Clichy, and Saint-Denis. (Or, as one historian speculated, would the French Revolution have occurred if the Constituent Assembly had met in Dijon—rather than in Versailles, less than 20 miles from Paris?)

With the disappearance of insulated space, violence has become an everyday reality. The ecological lines within the cities have been breached and crime has spilled over into every neighborhood. In the ordinary experiences of everyday life, a middle-class child today is no more safe from assault than a working-class child was 25 years ago. More important, given the turmoil that is likely to develop in the next 25 years, we may see Washington become a hotbed of overt, mobilized political conflict. The problem of security has become immediate to our lives.

What is true domestically is, of course, true in the international sphere as well. John von Neumann once remarked that World War II was the last war of the old geopolitical strategists, who could count on space as the critical variable. In World War II, Russia still had an effective land mass into which it could retreat, even when Moscow was threatened by foreign armies. Today, in an age of intercontinental ballistic missiles, there are no hiding spaces in any part of the globe. And with large aircraft, isolated cities like Berlin could be saved by airlifts; or, as in the cases of the Congo, the Middle East, and Vietnam, vast supplies and whole armies could be transported 10,000 miles in short spans of time. The first act in city planning, Aristotle remarks in the *Politics*, is the building of the city's walls, for a city without walls is an invitation to invasion. If space and security meant "freedom to go about one's affairs" and a relaxed social fabric, then the freedom and relaxation that America has known for a hundred years may be at an end.

Economic abundance

The United States, as the late David Potter remarked, was a "people of plenty." It was not just the fertile soil, the large forests, the vast seams of coal, the large veins of iron ore, and the lake-and-river system that could tie them together—though all of these were essential. America's primary bounty was the ingenuity, energy, and character of its people. Long before industrialization, in the 1840's, visitors to this country remarked on the kind of production and the modes of social organization that permitted the United States to take the lead in the manufacture of goods. There was, for example, Oliver Evans' continuous flour-milling system, which showed the way for the coordinated packing-house slaughter of animals and later for the assembly line of Henry Ford. They were symbolized by Eli Whitney's invention of simple templates, so that untutored mechanics could draw and cut a standardized part, which in time led to the mass production of cheap watches and hundreds of other consumer items.

Previously, as Brooks Adams observed, economic power had depended on access to metals and the strategic control of trade routes. But the United States had led the way to economic power through its supremacy in applied science and the new arts of management.

The central question is whether the United States can maintain, if it has not already lost, this supremacy. In a familiar principle of economic development, a nation arriving "later" not only has an advantage in being able to use the more advanced technology but also is not burdened by the huge depreciation costs of the older technology, and can thus leapfrog ahead of the initial innovators —a theme that Thorsten Veblen developed in his book *Imperial Germany*. There is a similar point in Raymond Vernon's thesis of the "product cycle": As a product becomes standardized in its use, other countries can reap production savings in labor and other costs so that, as in textiles, typewriters, or radios, production moves from the more advanced to the less developed country. To this extent, the United States, like England at the turn of the century, is caught in the turn of the economic product cycle and is losing its initial gains. It has even been suggested by the economic historian Charles P. Kindleberger that the United States may now have reached its "economic climacteric."

The areas of American economic "advantage" today form an odd mixture: food, military weapons, aircraft, computers, and a broad area of highly advanced technology comprising "miniaturization" (i.e., such semi-conductors as transistors and micro-processing) and certain optical processes (e.g., lasers). Yet most of these advantages are highly contingent. The United States is now a major food-exporting country, but its continuing advantage rests on uncertain climatic and political factors, such as the future ability of the Soviet Union and the Southeast Asian countries to overcome their agricultural deficits. Large amounts of military weapons now go to client states, but this is primarily a political rather than an economic factor. Miniaturization and optical technology were quickly mastered by Japan, and it is questionable how long our consistent lead will be maintained. Only in computers and aircraft is there a stable lead.

Yet the crucial fact is not these particular advantages for the balance of trade and payments, but a major change in the character of corporate income. Though foreign trade, given the size of this country and the magnitude of its economic activity, is still under 10 per cent of GNP, about 20 per cent of all corporate earnings comes from overseas. In this respect, two issues will become enormously important in the next decade. One is the fact that such countries as Germany and Japan are beginning to approach the limit of their advantage in the product cycle and in the export of goods, and a massive restructuring of their economies is taking place, one in which "know-how" and capital, competitive with the United States, are becoming the largest exports. And the second fact is that the United States, with its increasing dependence on overseas sales and investments for corporate earnings, becomes more and more dependent on the political conditions of those countries.

American economic abundance is now tied inextricably to the world economy, at a time when the United States is less able to enforce its economic or political will on other nations. Given the scale of American corporate investment abroad, the United States may in the next decade become a *rentier* economy, its margin of abundance dependent on the earnings of those overseas investments. And there is a major political question whether the less developed countries would allow such a *rentier* arrangement to remain.

To all this must be added the more familiar domestic problems of the growth of services and the rise of entitlements. If economic abundance begins to shrink, the main question is whether the majority of Americans will accept increased tax burdens and the reduction of private consumption as the price of economic and social redress. And if they do not, will the poor accept this extraordinary reversal? In the decade to come, this will be a potential source of serious discord in the country.

The two-party system

Richard Hofstadter has written, apropos of Louis Hartz's *The Liberal Tradition:* "One misses . . . in a book that deals with what is uniquely American two of our vital unique characteristics: our peculiar variant of federalism and our two-party system. Without a focus on federalism, we are tempted to downgrade the inventiveness of the American political system—for we were the pioneers in the development of the modern popular party and of the system of two-party opposition—but we miss the chance to see how conflict was both channeled and blunted in American history." The party system in the United States—which many persons take to be a unique institution to constrain conflict—was unforeseen at the beginning of the Republic. There is no mention of parties in the Constitution. In fact, to the degree that parties were discussed, their existence was deplored as partisan and as polarizing the society. In contrary fact, however, the American party system has limited the polarization of issues and forced the very compromises that are anathema to partisan politics. It is that fact which makes the present decomposition of the party system so troublesome when considering the future of American politics.

Politics in the United States has not been non-ideological. As many shades of ideology have been present in the United States as there are colors in the spectrum. What has been different in the United States is the fact that single ideological and class divisions, except for slavery, could not divide the polity along a single unyielding dimension. (And slavery could do so because it was concentrated in a single region.) In the nature of the multiple claims mediated by the political system, partisans of different ideologies had to compromise their demands or work only as single-issue groups within the larger framework. Thus when George Henry Evans sought to promote the Homestead Act in order to provide free land as a solution for labor ills, he did not, contrary to earlier impulses, start a new party, but worked within Congress to get the support of individuals from different parties on that issue alone. And when Samuel Gompers put the American Federation of Labor into politics in the 1890's, he angered the Socialists (who at that time had come close to capturing the leadership of that organization) by proclaiming the slogan, "Reward your friends, punish your enemies." How else, he explained, could one win remedial legislation, if one did not support those who had introduced and worked for that legislation? In the United States, because of the party system, ideology had shrunk to issues.

Along with the two-party system, different axes of social division weakened ideological politics in American life, and also the shifting emphases, at different historical periods, of different sociological divisions. Along one axis there have been economic and class issues which divided farmer and banker, worker and employer, and led to the functional and interest-group conflicts that were especially sharp in the 1930's. Along a different axis were status-group conflicts—the politics of the 1920's, and to some extent those of the 1950's, with the rural small-town Protestant intent on defending his "traditionalist" values against the cosmopolitan, urban liberal seeking to install new "modern" values. The McCarthyism of the 1950's was an effort by traditionalist forces—Joseph McCarthy's strongest support came from small businessmen—to impose a uniform political morality on the society by conformity to a single definition of ideological Americanism. In contrary fashion, the McGovern campaign of 1972 was fueled largely by a "new politics" which represented the most radical tendencies of the modernists —women's lib, sexual nonconformists, and cultural radicals in an alliance, for the moment, with black and other ethnic minority groups.

The importance of these two axes is that divisions along economic lines have not been congruent with cultural divisions. The labor movement in the United States, which has been consistently Democratic, is actively hostile to cultural radicalism. Farmers and small businessmen, who are usually Republican, cross the party line in times of economic crisis. At different historical periods, the economic or the status issues have been salient, and thus it has been difficult to maintain the historical continuity of groups on ideolog-

ical issues. The unique vitality of the American party system was
to maintain a shifting balance between different social forces, and
when there was too great a disequilibrium, realignments took place,
as they have about five times in American political history.

Today it seems likely that the party system in the United States
is in disarray, if not in complete deterioration.[15] Walter Dean Burnham, an unusually keen analyst, has in fact argued as follows:

> The American electorate is now deep into the most sweeping behav-
> ioral transformation since the Civil War. It is in the midst of a critical
> realignment of a radically different kind from all others in American
> electoral history. This critical realignment, instead of being channeled
> through partisan voting behavior as in the past, is cutting across older
> partisan linkages between rulers and ruled. The direct consequence of
> this is an astonishingly rapid dissolution of the political party as an
> effective "guide" or intervenor between the voter and the objects of his
> vote at the polls. . . . This is a realignment whose essence is the end of
> two-party politics in the traditional understanding; in short, it is a
> *caesura* in American political evolution, a moment in time at which we
> close a very long volume of history and open a brand-new one.[16]

The relevant evidence can be quickly summarized. First is the
decline of party identification, and the rise of the politically independent voter. Second is the fact that the rise in independence is
concentrated almost entirely among the young. Persons over 40
were virtually undisturbed in their political allegiances by the turmoil of the 1960's. But 26 per cent of the voters who were in their
20's in the 1960's registered as independents, and contrary to pre-

---

[15] My comments on this development can be brief: It is discussed at length in
Samuel P. Huntington's article in this issue, "The Democratic Distemper,"
while the special role of the American party system in moderating conflict is
discussed in S. M. Lipset's article, "The Paradox of American Politics."

The distinction between cultural-status and economic axes of politics is elu-
cidated in the essays by Hofstadter, Lipset, and myself in *The Radical Right*
(New York, Doubleday, 1963) and pursued historically by Joseph Gusfield in
Symbolic Crusade: Status Politics and the American Temperance Movement
(Urbana, University of Illinois Press, 1963).

[16] Walter Dean Burnham, "American Politics in the 1970s: Beyond Party?" (Unpublished paper). For other sources which mobilize data on these questions,
see James L. Sundquist, "Whither the American Party System," *Political Science Quarterly* (December 1973); Paul R. Abramson, "Generational Change in
American Electoral Behavior," *American Political Science Review* (March 1974);
Paul R. Abramson, "Why the Democrats Are No Longer the Majority Party,"
(paper prepared for the American Political Science Association, September
1973); and Arthur Miller *et al.*, "A Majority Party in Disarray," (paper prepared for the American Political Science Association, September 1973).

For historical data as background, see James L. Sundquist, *Dynamics of the
Party System* (Washington, D.C., Brookings Institution, 1973) and Walter Dean
Burnham, *Critical Elections and the Mainsprings of American Politics* (New
York, W. W. Norton, 1970).

vious experience, in which individuals identify with parties as they grow older, the proportion of independents in that age cohort had risen to 40 per cent 10 years later. The major result of all this has been a startling rise in "ticket-splitting" between the Presidential and Congressional contests, from 11.2 per cent in 1944 to 44.1 per cent in 1972.

The party machines themselves have largely broken down. The rise of public welfare and the growth of public unionism had already substantially reduced the role of patronage in supporting the party machines. Now the revolution in political campaign techniques, primarily the emergence of television as the principal channel of communication between candidate and voter, has robbed the party of one of its basic functions—the organization and management of campaigns.[17]

Issue politics

All this has gone hand in hand with a more troubling change in American politics—the swift rise of single, salient issues which have tended to polarize the electorate sharply. As party identification has decreased, individuals have focused their political identities on specific issues which symbolize their grievances and concerns about the society. The various readings of the Michigan Survey Research Center show an increasing issue-consciousness and issue-intensity among the electorate in the 1960's. In that decade, this was centered, by and large, on three issues: Vietnam, "race," and a cluster of concerns that involved drugs, youth rebellion, street crime, "coddling" of criminals, "permissiveness," and the like, which can generally be labeled "cultural." On the whole, these were not economic-class issues, and as a result it was evident that the old liberal coalition that had been built by the New Deal was falling apart.

---

[17] If one is to believe some recent arguments by political scientists, "Elections are now waged through the mass media which have supplanted political parties as the major intermediary between office seekers and the electorate...." (Thomas E. Patterson and Ronald P. Abeles, "Mass Communication and the 1976 Presidential Election," *Items,* published by the Social Science Research Council, June 1975, p. 13.) This is a sweeping claim, indeed. The "received knowledge" in the field has been skeptical about the powers of the mass media. The standard work—*Personal Influence,* by Elihu Katz and Paul Lazarsfeld (Free Press of Glencoe, 1955)—argued that the mass media serve largely to reinforce existing attitudes or to give individuals a "language" to express ideas, whereas actual influence is a two-step process in which "gatekeepers" or "style leaders" shape the attitudes and tones of small groups of followers who take their cues from these "influentials." If in 20 years there has indeed been a change in the patterns of influence, it is a major change in behavioral patterns.

In the past, when such massive shifts have taken place, they have set the stage for a "critical realignment." The "present" party structure came into being in the 1930's, during the Depression, when millions of voters made a permanent change in party identification, the country's previous normal Republican majority having been established in the critical election of 1896. Another "critical realignment" has since been expected by both the "right" and the "left."

And yet it does not seem as if any "critical realignment" will actually take place. For one thing, the new economic issues of the 1970's cut sharply across the older social issues. There is the dual problem of inflation and unemployment. But what is the specific "conservative" or "liberal" response? What characterized the New Deal was the commitment to government activism and intervention, as against that of the older Republicans, who feared and fought *any* government policy. But *every* administration is "activist" today. Nixon wanted "market" solutions, but established wage-and-price controls. Ford wanted to reduce government spending but reversed himself to create the largest budget deficit in American economic history (as did Eisenhower in 1958, when unemployment began to rise). One has to distinguish rhetoric from the political imperatives: The fact is that no administration today can escape the need for state management of the level of economic activity.

The more troublesome consideration is the increase in the general distrust by many individuals of the political system itself. In 1973, the loss of confidence in government and institutions reached majority proportions, according to the Louis Harris poll for a Senate committee. What is striking is how generalized and widespread this discontent has become. Almost all sociological analyses of politics start from standard demographic variables such as race, religion, region, income, education, and age, and relate political attitudes to social class clusters. It has been assumed that alienation fluctuates more in some demographic groups than in others. But some recent analyses of political alienation from 1952 to 1968 suggest a startling lack of correspondence between demographic status and ideological attitudes; the growing sense of alienation in this period would seem to be equal among all groups.[18]

---

[18] See James S. House and William M. Mason, "Political Alienation in America, 1952-1968," *American Sociological Review* (April 1975). Paul R. Abramson writes: "The persistent relationship of social class to partisan choice is one of the most extensively documented facts of American political life. . . . But the economic, social, and political changes of postwar America have eroded the relationship between social class and partisan choice." ("Generational Change

In the past, most of the partisan issues in American life have been converted into interest-group issues, in which particular advantages could be specified, so that deals and trade-offs could mediate differences. But more of the issues today—especially the symbolic ones —resist such compromise: They tend to be all-or-nothing, rather than more-or-less. When such symbolic issues as Vietnam or race become salient, the intensity of partisan feelings grows, and individuals are more ready to resort to extra-parliamentary, extra-legal means, or street violence, to express their views. And when such issues multiply, the level of generalized distrust of the system rises, and individuals tend to support extremist leaders—who, in this country, are mainly on the right.

A democratic society has to provide a mode of consistent representation of relatively stable alignments, or modes of compromise, in its polity. The mechanism of the American polity has been the two-party system: If the party system, with its enforced mode of compromise, gives way, and "issue politics" begins to polarize groups, we have then the classic recipe for what political scientists call "a crisis of the regime," if not a crisis of disintegration and revolution. Few would claim that this is an immediate possibility, but the point is that a structural strain has been introduced into the society and that a major element in the social stability of the country —the meaning of American exceptionalism—has been weakened. That is the danger before us.

IV. Constitutionalism and comity

In any root discussion of American society, we have to return to political philosophy. The American political system at its founding was a philosophical response to (and, in turn, creatively shaped) the social structures of 18th- and 19th-century America. There were two distinguishing features: First, the American Revolution, unlike the French, was primarily a *political*, not a *social*, revolution. It sought to provide self-government and individual freedom and it

in American Electoral Behavior," *American Political Science Review,* March 1974) For additional data, see Arthur Miller *et al., op. cit.*

Given the greater salience of culture as a motivation for individuals—in shaping a life-style and expressing themselves politically—it was inevitable, perhaps, that there would be a discordance between demographic statuses and behavior. In my essay, "The Cultural Contradictions of Capitalism" (*The Public Interest,* No. 21, Fall 1970), I argued that there had emerged a greater latitude for "discretionary social behavior" (which paralleled the economic idea of "discretionary income"), and that the standard sociological variables based on demographic attributes were no longer reliable predictors of life-style or "indiscretionary" social behavior.

assumed that any social changes would take place *outside* the political arena, by individuals freely shaping their own lives. It sought
to emancipate civil society from the state. To that extent it was the
classic *bourgeois* or liberal revolution, made easier by the absence
of settled feudal institutions; what was overthrown was political
authority 3,000 miles away. Second, the Revolution established a
constitutional structure of governance. A framework of powers was
laid out whose scale and institutions derived from an agrarian and
mercantile society, but whose principles were drawn from an older
font of wisdom—the classical view of politics which knew the threat
of tyranny that derives from the demagogic manipulation of the
masses and the centralization of power in a single set of institutions.
America was exceptional in being, perhaps, the only fully bourgeois-
liberal polity. Its sociological foundation was the denial of the pri-
macy of politics for everyday life.

Almost from the start, however, or at least from the 1830's and
1840's, the effort to create a *social* revolution began to transform
the political system. Government was to be used for social purposes, i.e., redistributive and redressive policies. The adaptive tasks
of American society in the last 150 years have been the creation of
new institutions to reconcile political power—its inherent corruption and misuse, and also its capability, through law or command,
to mobilize resources for common ends—with the new demands
created by economic development, changes in the occupational and
class character of the society, and the need for redress. In sequence,
we have seen the assumption of judicial review of legislative and
executive decisions, the creation of regulatory and administrative
agencies, themselves possessed of quasi-judicial authority, and the
establishment of a social welfare state. All of this took place within
the commitment to constitutionalism.

The problem which the nation faces in the coming decades is
how to maintain the framework of constitutionalism in mediating
the multiple conflicting demands that are upon us now and that
will multiply in the next decades—since the "social" and the "political" are now so inextricably joined. The liberal theory of society
was that law should be formally rational, i.e., procedural and not
substantive; that government was to be an umpire, or at worst a
broker, and not an intervening force in its own right. Yet in every
way the decisions of government today—from taxes to purchases,
from regulation to subsidies, from transfer payments to services—
are active forms of intervention whose consequence, if not direct
intention, is redress: a set of actions that antagonizes the losers yet

satisfies the gainers only grudgingly, since no one ever gets his full claims, nor acknowledges his gains as being enough. We have few principles in political philosophy and public law to justify a collective society or to establish a consistent principle of redress. We have few ideas—and this is the challenge to economists and social scientists—on how to use market and decentralized mechanisms for communal ends. Our resources, physical, financial, and intellectual, are strained.

If constitutionalism—the common respect for the framework of law, and the acceptance of outcomes under due process—fails, or is rejected by significant sections of the society, then the entire framework of American society would collapse as well. It is in this sense that the last remaining "exceptionalism" must persist.

The recognition of history

The shaping elements of any society, as I said earlier, are nature, religion, and history. The United States began with no "history," —the first such experiment in political sociology—and for much of its existence as a society, its orientation was to the "future," to its Manifest Destiny and mission. Today that sense of destiny has been shattered. Nature and religion have vanished as well. We are a nation like all other nations—Santayana once said that Americans were inexperienced in poisons, but we have acquired skill in that area as well—except that we have, *in looking back,* a unique history, a history of constitutionalism and comity.[19] We have been a society that has, by and large, maintained a respect for individual rights and liberties: The idea of being a "free people" has not been traduced, the principles of due process and law have remained inviolate. For all the domestic ills or foreign "crimes" of the United States, its record as a civilized society commands respect—especial-

---

[19] The idea of "comity" comes from Richard Hofstadter, who, in the reflective, concluding sections of his *The Progressive Historians* wrote:

> Finally, there is a subtler, more intangible, but vital kind of moral consensus that I would call comity. Comity exists in a society to the degree that those enlisted in its contending interests have a basic minimal regard for each other: one party or interest seeks the defeat of an opposing interest on matters of policy, but at the same time seeks to avoid crushing the opposition, denying the legitimacy of its existence or its values, or inflicting upon it extreme and gratuitous humiliations beyond the substance of the gains that are being sought. The basic humanity is not forgotten; civility is not abandoned; the sense that a community life must be carried on after the acerbic issues of the moment have been fought over and won is seldom far out of mind; an awareness that the opposition will someday be the government is always present.

ly compared to the savageries of the Soviet Union or Germany, or the newer states of Rwanda, Burundi, or Uganda—and we need not be apologetic on that score.

It has been said that there is a decay of legitimacy in the country and that this is a source of the potential disintegration of the nation. But this observation fails to make a necessary distinction between a *regime* and a *society*. A government, as Edmund Burke insisted long ago, is a contrivance, an instrument to deal with wants. But a society is a people shaped by history and bound by comity. It is the rupture of comity, the play of ideological passions to their utmost extreme, that shreds the society and turns the city into a holocaust.

Some conditions that have constrained conflict—the character of the party system—have been weakened. The recent political history of the successive administrations has left the nation with much moral disrepute. All of this places a great responsibility on the leadership of the society. This necessitates the recreation of a moral credibility whose essential condition is simple honesty and openness. It means the conscious commitment in foreign policy to limit national power to purposes proportionate with national interests and to forego any hegemonic dream, even of being the moral policeman of the world. Domestically it means the renewed commitment to the policy of inclusion whereby disadvantaged groups have priority in social policy, both as an act of justice and to defuse social tensions that could explode. The act of "conscious will" has to replace the wavering supports of American exceptionalism as the means of holding the society together.

Of all the gifts bestowed on this country at its founding, the one that alone remains as the element of American exceptionalism is the constitutional system, with a comity that has been undergirded by history. And it is the recognition of history, now that the future has receded, which provides the meaning of becoming twice-born. America was the exemplary once-born nation, the land of sky-blue optimism in which the traditional ills of civilization were, as Emerson once said, merely the measles and whooping cough of growing up. The act of becoming twice-born, the entrance into maturity, is the recognition of the mortality of countries within the time scales of history.

History, as Richard Hofstadter observed eloquently in the concluding pages of the book which took the measure of the Progressive historians, provides "not only a keener sense of the structural complexity of our society in the past, but also a sense of the moral complexity of social action." For this reason, history has always disturbed the radical activists, who fear that the sense of complexity leads to political immobility since, as Hofstadter remarks, "history does seem inconsistent with the coarser rallying cries of politics."

And yet, history does provide us with a double consciousness of the need for reflection and also commitment. "As practiced by mature minds," Hofstadter concludes, "history forces us to be aware not only of complexity but of defeat and failure: It tends to deny that high sense of expectation, that hope of ultimate and glorious triumph, that sustains good combatants. There may be comfort in it still. In an age when so much of our literature is infused with nihilism, and other social disciplines are driven toward narrow positivistic inquiry, history may yet remain the most humanizing among the arts." And if the United States, as a polity, remains aware of the moral complexity of history, it may also remain humanized among the nations.